T0250280

A Simple Introduction to Python

A Simple Introduction to Python is aimed at pre-university students and complete novices to programming. The whole book has been created using Jupyter notebooks. After introducing Python as a powerful calculator, simple programming constructs are covered, and the NumPy, MatPlotLib and SymPy modules (libraries) are introduced. Python is then used for Mathematics, Cryptography, Artificial Intelligence, Data Science and Object Oriented Programming. The reader is shown how to program using the integrated development environments: Python IDLE, Spyder, Jupyter notebooks, and through cloud computing with Google Colab.

Features:
- No prior experience in programming is required.
- Demonstrates how to format Jupyter notebooks for publication on the Web.
- Full solutions to exercises are available as a Jupyter notebook on the Web.
- All Jupyter notebook solution files can be downloaded through GitHub.

GitHub Repository of Data Files and a Jupyter Solution notebook:

https://github.com/proflynch/A-Simple-Introduction-to-Python

Jupyter Solution notebook web page:

https://drstephenlynch.github.io/webpages/A-Simple-Introduction-to-Python-Solutions.html

In 2022, **Stephen Lynch** was named a National Teaching Fellow, which celebrates and recognizes individuals who have made an outstanding impact on student outcomes and teaching in higher education. He won the award for his work in programming in STEM subjects, research feeding into teaching, and widening participation (using experiential and object-based learning). Although educated as a pure mathematician, Stephen's many interests now include applied mathematics, cell biology, electrical engineering, computing, neural networks, nonlinear optics and binary oscillator computing, which he co-invented with a colleague. He has authored 2 international patents for inventions, 8 books, 4 book chapters, over 45 journal articles, and a few conference proceedings. Stephen is a Fellow of the Institute of Mathematics and Its Applications (FIMA) and a Senior Fellow of the Higher Education Academy (SFHEA). He is currently a Reader with MMU and was an Associate Lecturer with the Open University from 2008 to 2012. In 2010, Stephen volunteered as a STEM Ambassador, in 2012, he was awarded MMU Public Engagement Champion status, and in 2014, he became a Speaker for Schools. He runs national workshops on "Python for A-Level Mathematics and Beyond" and international workshops on "Python for Scientific Computing and TensorFlow for Artificial Intelligence." He has run workshops in China, Malaysia, Singapore, Saudi Arabia, and the USA.

Chapman & Hall/CRC
The Python Series

About the Series

Python has been ranked as the most popular programming language, and it is widely used in education and industry. This book series will offer a wide range of books on Python for students and professionals. Titles in the series will help users learn the language at an introductory and advanced level, and explore its many applications in data science, AI, and machine learning. Series titles can also be supplemented with Jupyter notebooks.

Image Processing and Acquisition using Python, Second Edition
Ravishankar Chityala, Sridevi Pudipeddi

Python Packages
Tomas Beuzen and Tiffany-Anne Timbers

Statistics and Data Visualisation with Python
Jesús Rogel-Salazar

Introduction to Python for Humanists
William J.B. Mattingly

Python for Scientific Computation and Artificial Intelligence
Stephen Lynch

Learning Professional Python Volume 1: The Basics
Usharani Bhimavarapu and Jude D. Hemanth

Learning Professional Python Volume 2: Advanced
Usharani Bhimavarapu and Jude D. Hemanth

Learning Advanced Python from Open Source Projects
Rongpeng Li

Foundations of Data Science with Python
John Mark Shea

Data Mining with Python: Theory, Applications, and Case Studies
Di Wu

A Simple Introduction to Python
Stephen Lynch

For more information about this series please visit: https://www.crcpress.com/Chapman--HallCRC/book-series/PYTH

A Simple Introduction to Python

Stephen Lynch
Manchester Metropolitan University,
United Kingdom

CRC Press
Taylor & Francis Group
Boca Raton London New York

CRC Press is an imprint of the
Taylor & Francis Group, an **informa** business
A CHAPMAN & HALL BOOK

Designed cover image: Python Software Foundation

First edition published 2024
by CRC Press
2385 NW Executive Center Drive, Suite 320, Boca Raton FL 33431

and by CRC Press
4 Park Square, Milton Park, Abingdon, Oxon, OX14 4RN

CRC Press is an imprint of Taylor & Francis Group, LLC

© 2024 Stephen Lynch

Library of Congress Cataloging-in-Publication Data

Names: Lynch, Stephen, 1964- author.
Title: A simple introduction to Python / Stephen Lynch, Manchester Metropolitan University, United Kingdom.
Description: First edition. | Boca Raton : C&H, CRC Press, 2024. |
Series: Chapman & Hall/CRC the Python series | Includes bibliographical references and index.
Identifiers: LCCN 2023056099 (print) | LCCN 2023056100 (ebook) | ISBN 9781032751672 (hbk) | ISBN 9781032750293 (pbk) | ISBN 9781003472759 (ebk)
Subjects: LCSH: Python (Computer program language) | Computer programming.
Classification: LCC QA76.73.P98 L963 2024 (print) | LCC QA76.73.P98 (ebook) | DDC 005.13/3--dc23/eng/20240212 LC record available at https://lccn.loc.gov/2023056099 LC
ebook record available at https://lccn.loc.gov/2023056100

ISBN: 978-1-032-75167-2 (hbk)
ISBN: 978-1-032-75029-3 (pbk)
ISBN: 978-1-003-47275-9 (ebk)

DOI: 10.1201/9781003472759

Typeset in Latin Modern font
by KnowledgeWorks Global Ltd.

Publisher's note: This book has been prepared from camera-ready copy provided by the authors.

To my family,
my brother Mark and my sister Jacqueline,
my wife Gaynor,
and our children, Sebastian and Thalia,
for their continuing love, inspiration and support.

Contents

Preface

Currently, Python is the most popular programming language in the world. It is open source and completely free to the user. It is also fun to program with and extremely powerful at solving real-world problems.

This book is for pre-university students or complete novices to programming. This is the third book I have written on Python. The most recent book, **Python for Scientific Computing and Artificial Intelligence** [1], published by CRC Press in 2023, is aimed at high-school students, undergraduates, postgraduates and scientists, and concentrates on interdisciplinary science, computational modeling, simulation and programming. The more advanced book, **Dynamical Systems with Applications using Python** [2], published by Springer International Publishing, is set at a final-year undergraduate and postgraduate level and covers the mathematical theory and research in some detail. This book is a precursor to both [1] and [2] and provides a gentle introduction to programming in Python and is aimed at complete novices.

In the second part of the Preface, the reader is shown three (from many) methods to access Python. The Python Integrated Development Learning Environment (IDLE) is suitable for Chapters 1 to 3. For Scientific Computing and the material in Chapters 4 to 10, the reader will need to use Jupyter notebooks or Spyder through Anaconda, or Google Colab, via cloud computing.

Each chapter provides both examples and exercises. In order to understand programming, the reader must attempt these exercises. Learning to program is like learning to ride a bike – you have to fall off many times before mastering the art. Expect to make many mistakes, learn from those mistakes and then move on to the next chapter. Full working solutions to all of the exercises and other resources will be provided via GitHub, where readers will be able to download all files for free. There will also be a web page with full-worked solutions, where readers can simply copy and paste code and run the programs in a Python environment. Readers should note that Python programs can easily be generated with ChatGPT (Chat Generative Pre-trained Transformer), developed by OpenAI, and other alternatives such as Microsoft Bing, Perplexity AI and Google Bard AI, for example. Readers should ask these AI chatbots why humans should learn how to program – they give some very sound arguments.

Chapter 1 shows the reader how Python can be used as a simple, powerful calculator, and the first library (or module) is introduced. The Math library consists of functions (written in Python) which can be called within Python. Some of the functions will be familiar to most readers (asin, sin, exp, gcd, lcm, sqrt, etc), and some unfamiliar functions (ceil, floor, fmod, radians, trunc) will be defined here. Chapter 2 starts with lists, tuples, sets and dictionaries and then moves on to simple programming, defining functions (think of adding buttons to your Python calculator), loops and conditional constructs (if, elif, else). In Chapter 3, the turtle library is loaded into a notebook and simple fractals (images repeated on ever-reduced scales) are plotted using recursive functions and iteration. Numerical Python (NumPy) and the Matrix Plotting Library (MatPLotLib) are used in Chapter 4 when dealing with arrays, matrices, vectors, tensors and plots. Cloud computing with Google Colab (Collaboratory) is covered in Section 5.1, where the reader can use Jupyter notebooks without requiring any software on their own computer. The next section introduces the reader to formatting notebooks, inserting titles, figures and mathematical equations using LaTeX. Next, the Symbolic Python (SymPy) library for symbolic computation is introduced and the final section shows the reader how to access GitHub, the AI-powered developer platform to build, scale, and deliver secure software. As well as storing and sharing Python programs here, the reader should know that it also provides a platform to publish your own web pages – as the author has done for his personal web pages and the solutions to this book.

Chapter 6 shows the reader how Python can help understand basic Mathematics. Using Python can help provide a deeper understanding of the mathematics, I am a strong advocate of bringing it into all school curricula across the world. Not only can it help with the understanding but it can also be lots of fun too – something that is missing from a lot of children's mathematics education! Chapter 7 provides a basic introduction to cryptography and cyber security starting with some well-known historical ciphers and ending with a simple example of the Rivest-Shamir-Adleman (RSA) algorithm. Artificial Intelligence (AI) by way of artificial neural networks (ANNs) is addressed in Chapter 8. There are simple ANNs for the AND, OR and XOR gates from logic, and the backpropagation algorithm is explained using a very simple ANN. The well-known data set for Boston housing from the 1970s is discussed and a full working program of the Boston housing ANN is provided in the solutions notebook. As an introduction to Data Science, Chapter 9 starts with the Python and Data Analysis (PANDAS) library and then the reader is given examples of how to load, pre-process data, present the data visually, explore the data and analyze and communicate the results. Four large data sets (LDSs) are used in this chapter, and all LDSs can be downloaded through GitHub.

Chapter 10 provides a simple introduction to object oriented programming (OOP), which is a style of programming which is different to functional and procedural programming encountered in earlier chapters. The concept is based on objects which contain data (attributes) and code (methods). Using student, employee and vehicle objects, the concepts of encapsulation, inheritance and polymorphism are briefly explained. An example of a Brick Breaker Game OOP is listed in the exercises, and a full working program of the game is provided in the solution notebook.

Data files and full worked solutions to all of the chapter exercises will be available through GitHub:

`https://github.com/proflynch/A-Simple-Introduction-to-Python.`

Alternatively, readers can view a Jupyter notebook of all of the solutions here:

`https://drstephenlynch.github.io/webpages/A-Simple-Introduction-to-Python-Solutions.html.`

For applications in biology, chemistry, data science, economics, engineering, fractals and multifractals, image processing, numerical methods for ordinary and partial differential equations, physics, statistics and artificial intelligence (AI), see [1]. For more information on the theory and applications of continuous and discrete dynamical systems, the reader is directed to [2].

1. Lynch, S. (2023). Python for Scientific Computing and Artificial Intelligence. CRC Press, Boca Raton, FL, USA.

   `https://www.routledge.com/Python-for-Scientific-Computing-and-Artificial-Intelligence/Lynch/p/book/9781032258713#`

2. Lynch, S. (2018). Dynamical Systems with Applications using Python. Springer International Publishing, New York, USA.

http://www.springer.com/us/book/9783319781440

Accessing Python:

There are many Integrated Development Environments (IDEs) and code editors for Python. In 2023, the most popular include Atom, Dreamweaver, Eric, PyCharm, Pydev, Sublime Text, Thonny, Vim, Visual Studio Code and Wing, for example. The reader can find more information on each of these on the Web.

For this book, there are three easy ways to access Python through the internet. The first is using Python IDLE, which is popular in schools around the world. The second is via the Anaconda Navigator, where users have access to Jupyter notebooks and Spyder (Scientific PYthon Development EnviRonment). The third popular way to access Python is via cloud computing using Google Colab. Readers need a Google account to use Colab. There are many videos freely available online demonstrating how each of these media is used. Here, I give a brief introduction to each one.

1. The Python Integrated Development Learning Environment (IDLE): is suitable for beginner-level developers and school children. The URL to access Python IDLE is:

https://www.python.org.

Figure 1 shows the Python homepage where readers can download Python IDLE. After downloading Python IDLE, readers can click on the IDLE logo (see the left hand of Figure 2), and an IDLE Shell is loaded where readers can use Python as a calculator. For writing programs (as in Chapters 2 and 3), open an untitled Editor Window as you would for a Word document. The file can then be saved in the same way that you save files in Microsoft.

For scientific computing and the material covered in Chapters 4 to 10, the reader will need to either download Anaconda (a distribution of the Python and R programming languages for scientific computing that aims to simplify package management and deployment) or use cloud computing via Google Colab.

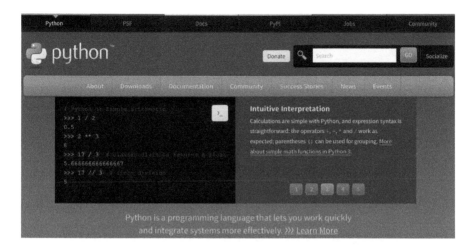

Figure 1   Homepage of Python.org. Readers can download Python IDLE here.

2. The URL to download Anaconda is:

https://www.anaconda.com/download.

Figure 3 shows the Anaconda distribution download page. Readers can download for Windows, Apple or Linux computers. Clicking on

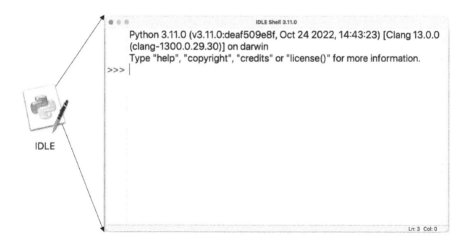

Figure 2   The Python IDLE Shell, where readers can use Python like a calculator. See Chapters 1 to 3.

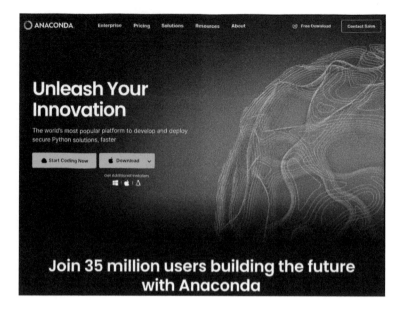

Figure 3 The homepage for Anaconda.

the Anaconda Navigator icon opens Anaconda Navigator window; see Figure 4, where readers have access to JupyterLab, Jupyter Notebook and Spyder. The other data science and statistics packages are not used in this book.

Figure 5 shows the Spyder environment in light mode on an Apple Mac. Those readers familiar with MATLAB® will notice similarities here. The Editor Window is where users can write Python programs. The top right window is where one can get help, explore variables, see plots and see files in folders. The Console Window is where Python can be used as a calculator.

Figure 6 is the window that opens on launching the Jupyter Notebook button. Jupyter Notebook only offers a very simple interface in which users can open notebooks, terminals and text files. JupyterLab offers a very interactive interface that includes notebooks, consoles, terminals, CSV editors, markdown editors, interactive maps and more.

Figure 7 is an untitled Jupyter Notebook. The reader can insert Markdown cells to insert titles, pictures and mathematical equations using LaTeX, or Code cells for Python code. To execute a cell hit the Run button or type **SHIFT+ENTER** in the cell. The reader can save the file or hit **File** and download as AsciiDoc, HTML (.html), LaTeX (.tex), Markdown, Notebook (.ipynb) or PDF via LaTeX. By downloading as

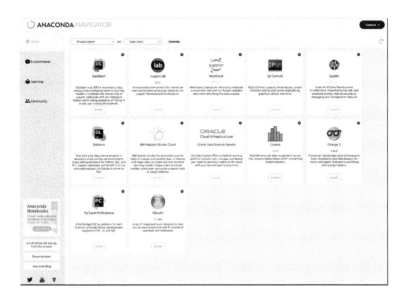

Figure 4   The Anaconda Navigator window, in this book we use Jupyter-Lab, Jupyter Notebooks and the Scientific Python Development Environment (Spyder) only.

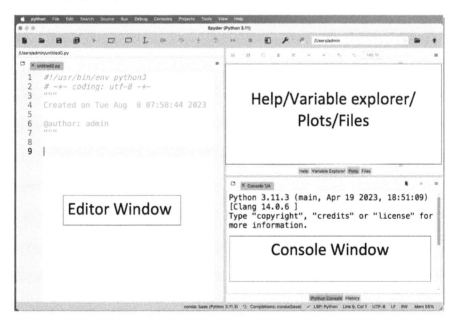

Figure 5   The Spyder environment in light mode on an Apple Mac.

Figure 6 The window that appears on launching **Jupyter Notebook**. Click on **New** and **Python 3 (ipykernel)** (see top right corner) to open a new Jupyter notebook.

html, the reader can publish the notebook on the Web. The author used GitHub to publish the solution notebook online.

3. Google Colab: To perform cloud computing and access Google Colab you need a Google account:

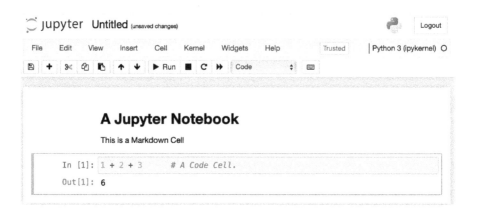

Figure 7 An untitled Jupyter notebook. The top cell is a Markdown cell. In [1]: is a code cell and Out[1]: shows the output when the cell is **Run**. One can also hit **SHIFT + ENTER** in the code cell.

https://colab.research.google.com/.

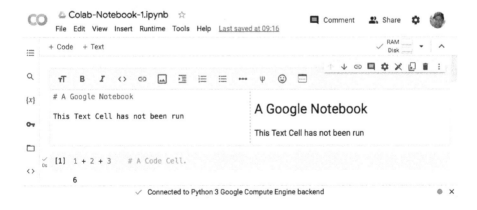

Figure 8 The window that opens for Google Colab. Create a new Jupyter notebook by clicking on **+ New notebook**, in the blue box.

Figure 8 shows the window that opens on launching Google Colab. This assumes that you have already logged in to your Google account. Simply hit on **New notebook** at the bottom of the page to obtain a Google Colab Jupyter Notebook, as seen in Figure 9.

In Figure 9, the Markdown cell has not been executed, but Colab shows what the output will be. Note that html uses the # symbol for

Figure 9 A Google Colab Jupyter notebook. The top cell is a Text cell and [1] is a Code cell.

Figure 10 The GitHub homepage.

titles. This notebook has been saved as Colab-Notebook-1.ipynb; see the top left corner. Colab notebooks can be saved directly to GitHub. See Chapter 5 for more details.

Now that we have suitable IDEs, the reader is ready to start using Python as a calculator and writing programs for more complex tasks.

Finally, the author recommends that the reader sign up to GitHub, a platform for hosting code that allows for version control and collaboration:

https://github.com.

Readers may use GitHub as free cloud storage for their projects and to conveniently host static websites. Again, there are plenty of free videos on the internet to show you how to do this.

## CHAPTER 1

Python as a Powerful Calculator

1.1 BODMAS

This section shows that Python follows the rules of BODMAS (Brackets, Order, Division, Multiplication, Addition, Subtraction).

Example 1.1.1. Compute the following:

(a) $4 + 5 - 6$;   (b) 2×3;   (c) $3 \times (2 - 5)$;

(d) $2 + 3 \times (6 + 2 \times 5) - 8$;   (e) $\frac{1}{2} \times \left(\frac{1}{3} - \frac{1}{4}\right)$.

The following solutions are computed using a Jupyter notebook. In [1]: is a code cell; if you hit **SHIFT + ENTER**, then the cell is executed and the result is Out[1]:.

DOI: 10.1201/9781003472759-1

```
In [1]: 4 + 5 - 6  # Python ignores anything after the hash.
Out[1]: 3
```

```
In [2]: 2 * 3       # Asterisk for multiplication.
Out[2]: 6
```

```
In [3]: 3 * (2 - 5)
Out[3]: -9
```

```
In [4]: 2 + 3 * (6 + 2 * 5) - 8
Out[4]: 42
```

```
In [5]: (1 / 2) * (1 / 3 - 1 / 4)  # Forward slash for divide.
Out[5]: 0.04166666666666666
```

1.2 FRACTIONS: SYMBOLIC COMPUTATION

To compute with fractions, without using decimals, readers must load the **Fraction** function from the **fractions** library.

The rules of fraction arithmetic are as follows:

1. $\frac{a}{b} \pm \frac{c}{d} = \frac{ad \pm bc}{bd}$, where $b, d \neq 0$;

2. $\frac{a}{b} \times \frac{c}{d} = \frac{a \times c}{b \times d}$, where $b, d \neq 0$;

3. $\frac{a}{b} \div \frac{c}{d} = \frac{a \times d}{b \times c}$, where $b, c, d \neq 0$.

The examples below are in the same Jupyter notebook as above, so start with the code cell In [6]:. To execute each cell, the reader hits SHIFT + ENTER.

Example 1.2.1. Compute the following, leaving the answers as fractions:

(a) $\frac{1}{2} + \frac{1}{3}$;   (b) $\frac{1}{2} \times 8$;   (c) $\frac{1}{3} \times \frac{2}{5}$;

(d) $\frac{1}{2} \div \frac{1}{3}$;   (e) $\frac{1}{2} \times \left(\frac{1}{3} - \frac{1}{4} \right)$.

The following solutions are computed using a Jupyter notebook. In [6]: is a code cell; if you hit **SHIFT + ENTER**, then the cell is executed. Note that In [6] has no output.

```
In [6]:  from fractions import Fraction
```

```
In [7]:  Fraction(1 , 2) + Fraction(1 , 3)
Out[7]:  Fraction(5, 6)
```

```
In [8]:  Fraction(1 , 2) * 8
Out[8]:  Fraction(4, 1)
```

```
In [9]:  Fraction(1 , 3) * Fraction(2 , 5)
Out[9]:  Fraction(2, 15)
```

```
In [10]:  Fraction(1 , 2) / Fraction(1 , 3)
Out[10]:  Fraction(3, 2)
```

```
In [11]:  Fraction(1 , 2) * (Fraction(1 , 3) - Fraction(1 , 4))
Out[11]:  Fraction(1, 24)
```

1.3 POWERS (EXPONENTIATION) AND ROOTS

To compute powers and roots, readers can import the **pow** and **sqrt** functions from the **math** library.

The rules of indices (powers) are as follows:

1. $x^a \times x^b = x^{a+b}$;

2. $x^{-a} = \frac{1}{x^a}$;

3. $x^a \div x^b = x^{a-b}$;

4. $(x^a)^b = x^{ab}$;

5. $x^{\frac{a}{b}} = \sqrt[b]{x^a}$.

Note that fractional powers result in roots.

Example 1.3.1. Compute the following:

(a) 2^8;   (b) $\left(\frac{1}{2}\right)^2$;   (c) 2^{-8};

(d) $\sqrt[3]{27} + \sqrt{25}$;   (e) $(2^3)^5 \left(\sqrt[4]{16} \times \sqrt[3]{\frac{1}{64}} + \sqrt{169} \right)$.

The following solutions are computed using a Jupyter notebook. In [12]: is a code cell; if you hit **SHIFT + ENTER**, then the cell is executed and the result is Out[12]:.

```
In [12]:  from math import * # Import all math functions.
          2**8                # Exponentiation.
Out[12]:  256

In [13]:  (1 / 2)**2
Out[13]:  0.25

In [14]:  pow(2 , - 8) # Power function.
Out[14]:  0.00390625

In [15]:  pow(27 , 1 / 3) + sqrt(25)
Out[15]:  8.0

In [16]:  (2**3)**5 * (16**(1 / 4) * pow(1 / 64 , 1 / 3) + sqrt(169))
Out[16]:  442368.0
```

Comments: Note that Python ignores anything typed after the hash (#) symbol. The comment symbol is extremely useful in all programming languages.

1.4 THE MATH LIBRARY (MODULE)

The math library includes the following functions:

acos(x), asin(x), asinh(x), atan(x), ceil(x), comb(n,k), cos(x), cosh(x), degrees(x), dist(x), erf(x), exp(x), fabs(x), factorial(x), floor(x), fmod(x,y), gcd(numbers), hypot(x,y), isnan(x), lcm(numbers), log(x), log10(x), radians(x), remainder(x,y), sin(x), sinh(x), sqrt(x), tan(x), tanh(x), trunc(x), e, inf, nan, pi, tau

To see definitions of these functions, type **help(math)** or follow this link:

https://docs.python.org/3/library/math.html.

Example 1.4.1. Use the math library to:

(a) compute e^2;

(b) use the **gcd** function to compute the greatest common divisor of 35 and 128;

(c) convert $180°$ to radians;

(d) compute $\sin\left(\frac{\pi}{2}\right)$;

(e) round π to ten decimal places.

The examples below are in the same Jupyter notebook as above, so start with the code cell In [17]:, in which all of the functions from the math library are imported, using the command **from math import ***.

```
In [17]: from math import * # Import all functions from the math library.

In [18]: exp(2)
Out[18]: 7.38905609893065

In [19]: gcd(35 , 128)
Out[19]: 1

In [20]: radians(180)
Out[20]: 3.141592653589793

In [21]: sin(pi / 2)
Out[21]: 1.0

In [22]: round(pi , 10)
Out[22]: 3.1415926536
```

The reader is encouraged to learn all of the functions from the math library. Examples can be found in the URL above.

Note that, for all exercises in this book, full-worked solutions can be viewed on the web.

https://drstephenlynch.github.io/webpages/A-Simple-Introduction-to-Python-Solutions.html.

EXERCISES

1. Compute the following:

   (a) $101 - 34 + 67$; (b) 12×7; (c) $4 \times (7 + 9 \times 3)$;
   (d) $2 - 2 \times (2 - 4)$; (e) $0.1 \div (0.6 - 0.05)$.

2. Compute the following fractions, leaving the answers as fractions:

   (a) $\frac{1}{4} - \frac{1}{5}$; (b) $\frac{2}{3} \times 30$; (c) $\frac{2}{5} \times \frac{5}{7}$;
   (d) $\frac{1}{3} \div 2$; (e) $\frac{1}{2} \times \left(\frac{1}{4} - \frac{1}{3}\right) \div \frac{1}{8}$.

3. Determine:

   (a) 2^{15}; (b) $\left(\frac{1}{3}\right)^3$; (c) 64^{-2};
   (d) $10^5 \times 10^{10}$; (e) $\left(2^5\right)^3 \left(\sqrt[3]{27} \times \sqrt[4]{625} - \sqrt[5]{32}\right)$.

4. Look up the math functions, fmod, factorial, log, floor, ceil and lcm. Use the math module to:

   (a) Compute fmod(36,5).
   (b) Compute 52!. How does this relate to a pack of playing cards?
   (c) Find $\ln(2)$, where $\ln(x) = \log_e(x)$ is the natural logarithm.
   (d) Determine floor(π)−ceil(e).
   (e) Compute the lowest common multiple of 6, 7 and 9.

## CHAPTER 2

Simple Programming with Python

2.1 LISTS, TUPLES, SETS AND DICTIONARIES

There are four built-in data types in Python, these are lists, tuples, sets and dictionaries. Each will now be defined:

A list is a collection of elements which is ordered and changeable, permitting duplicate elements. For example, $[1, 2, 3, 3, 4, 5]$, is a list with six elements. Note the square brackets and zero-based indexing is used.

DOI: 10.1201/9781003472759-2

A **tuple** is a collection of elements which is ordered and unchangeable, permitting duplicate elements. Tuples are usually used for heterogeneous (different) data types. For example, (4,5.42,6,6,"cherry","date"), is a tuple with six elements. Note the round brackets and zero-based indexing is used.

A **set** is a collection of elements which is unordered, unchangeable and unindexed with no duplicate elements. For example, {7,8,9,"fig","grape"}, is a set with five elements. Note the curly brackets.

A **dictionary** is a collection of key:value pairs which is ordered and changeable, without duplicate elements. For example, {"England" : "London" , "France" : "Paris" , "Germany" : "Berlin"}, is a dictionary of European capitals. Note the curly brackets and key:value pairs and zero-based indexing is used.

Python uses **zero-based indexing**, where the first element has index zero, the second element has index one and so on.

Slicing is used in Python to remove certain elements from data types.

The reader should execute the commands below and make up their own examples to get a deeper understanding.

Examples.

In/Out	Code Cells

In [1]:
```
# An empty list.
list_empty = []
print(list_empty)
```
Out[1]:   []

In [2]:
```
# Class type.
list1 = [1 , 2 , 3 , 4 , 5]
print(type(list1))
```
Out[2]:   <class 'list'>

In [3]:
```
# Zero-based indexing.
list1[0] , list1[1] , list1[-1]
```
Out[3]:   (1,2,5)

```
In [4]:    # Length, maximum and minimum.
           len(list1) , max(list1), min(list1)
Out[4]:    (6,5,1)
In [5]:    # Using range to generate numerical lists.
           list(range(5) , list(range(2 , 14 , 2))
Out[5]:    ([0,1,2,3,4], [2,4,6,8,10,12])
In [6]:    # Slicing lists.
           list1[1 : 3] , list1[2:] , list1[:1]
Out[6:]    ([2 , 3] , [3 , 3 , 4 , 5], [1])
In [7]:    # Lists of lists.
           list2 = [[1 , 2] , [3 , 4]]
           list2
Out[7]:    [[1 , 2] , [3 , 4]]
In [8]:    # The second element in the first list.
           list2[0][1]
Out[8]:    2
In [9]:    # Append and remove elements from a list.
           list1.append(10)
           print(list1)
Out[9]:    [1 , 2 , 3 , 3 , 4 , 5 , 10]
In [10]:   list1.remove(3) # Removes the first 3.
           print(list1)
Out[10]:   [1 , 2 , 3 , 4 , 5 , 10 ]
In [11]:   # Working with dictionaries.
           Capitals={"England":"London","France":"Paris"}
           Capitals.pop("England") # Remove item.
           print(Capitals)
Out[11]:   {"France":"Paris"}
In [12]:   Capitals.get("France")
Out[12]:   'Paris'
```

Consider In [2]:, above, the list1 is a list of five elements, $[1, 2, 3, 4, 5]$, list1$[0]$ is the first element, which is 1, list1$[-1]$, is the final element, which is 5. Slicing, list1$[1 : 3]$, starts at list1$[1]$, which is 2, and goes up to, but not including, list1$[3]$, which is 4. So list1$[1 : 3] = [2 , 3]$.

2.2 DEFINING FUNCTIONS (PROGRAMMING)

This section shows the reader how to define functions in Python. The reader should also run this program in Spyder and in a Jupyter notebook.

Example 2.2.1. Using Python IDLE, define a function that computes the area of a circle. Find the area of a circle of radius 4 cm. Note that Python also ignores text between triple quotes.

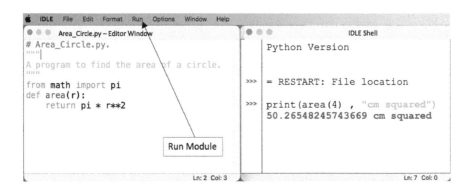

The area is computed to be 50.26548245743669 cm². Users have to run the module by hitting the **Run** button. Then, in the IDLE Shell window, one types **print(area(4) , "cm squared")** to get the answer. Note that in Python, programs need to be executed before they are called in the IDLE Shell.

Example 2.2.2. Using Python IDLE, define a function that computes the volume of a cylinder. Find the volume of a cylinder of height 5 cm and radius 4 cm.

The volume of the cylinder is computed to be 251.32741228718345 cm³. Remember to Run the module before calling the function.

Example 2.2.3. Using Python IDLE, define a function that converts degrees Centigrade to Kelvin. On execution of the program, input the temperature and output the result with eight place holders and four decimal places. Within formatting, the command {:08.4f} gives the output with eight place holders and four decimal places.

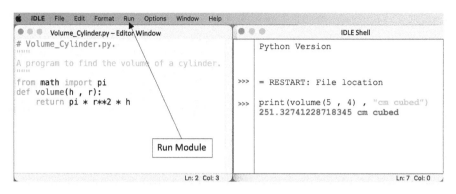

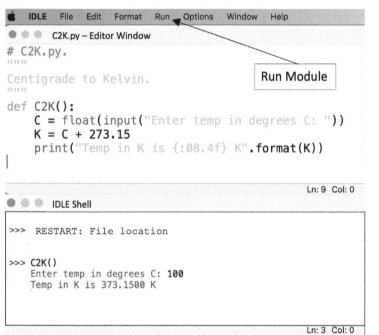

2.3 FOR AND WHILE LOOPS

We will be constructing examples using the following programming constructs:

for	**while**
for *item* in *object*:	while *expression*:
statement(s)	statement(s)

Note that the indentation levels in Python are vital to the successful running of the programs. When you hit the RETURN key after the colon, Python will automatically indent for you. These indentations can be lost when copying and pasting Python code.

Example 2.3.1. Within the Spyder interface, use a **for loop** to define a function that computes the first twenty terms of the Fibonacci sequence and output the sequence as a list.

The Spyder interface is shown below:

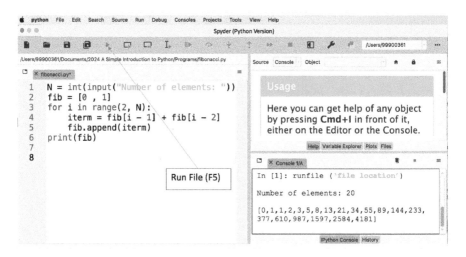

The first twenty terms of the Fibonacci sequence are displayed in the Console window in the lower right.

Preferences: By clicking on the **python** button (top left) readers can access Preferences, where options are available to the user, including setting the font size and working in light or dark mode. The Spyder interface above is shown in light mode.

Editor Window: This window is shown in the left column of the Spyder interface and is where users can write Python programs. Remember to **Run** the file. When defining functions, the file must be run before it is called in the Console window. Users can also **Undock** this window to get more room on their computer screen. **Redock**, will redock the window.

Help, Variable Explorer, Plots, Files, Window: The window in the top right corner is where users have access to help, can view variables in programs, view plots and save them, and see files in folders.

Console Window: The lower right window is where Python can be used as a calculator. Users can also **Undock** and **Redock** this window.

Example 2.3.2. Within Google Colab, use a **while loop** to define a function that computes the sum of the first n natural numbers.

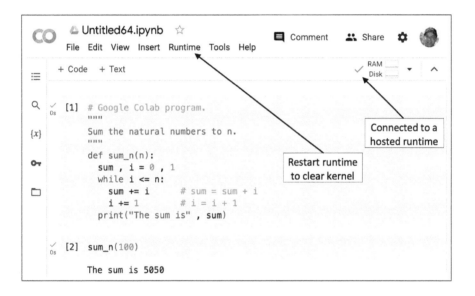

The sum of the first 100 natural numbers is 5050.

Google Colab: The URL for Google Colab is:

https://colab.research.google.com/.

Readers should note that this is cloud computing and you need access to the internet to use Colab. In the top right corner, the green tick next to the RAM and Disk icons indicates that you have **Connected to a hosted runtime**, in other words, you are connected to the Cloud. By clicking on **Runtime**, users can restart runtime and clear the kernel to start with [1] again. All of the variables stored will be lost. Programmers usually have to **Restart the Kernel** if a program hangs or something else goes wrong.

2.4 CONDITIONAL STATEMENTS, IF, ELIF, ELSE

We will be constructing an example using the following constructions

if, elif, else

if *expression*:
    body of if
elif *expression*:
    body of elif
else:
    body of else

Example 2.4.1. Within a Jupyter notebook, use if, elif, else commands to determine whether an integer is negative, positive or zero.

```
In [1]: # Save file as testinteger.py.
        # Run the Module (or type F5) in IDLE.
        """
        Test an integer.
        """
        def testint():
            testint=int(input("Enter integer: "))
            if testint < 0:
                print("Integer {} is negative".format(testint))
            elif testint > 0:
                print("Integer {} is positive".format(testint))
            else:
                print("The integer {} is zero".format(testint))
```

```
In [2]: testint()

        Enter integer: -296
        Integer -296 is negative
```

Note that the if, elif and else statements all end with a colon. Python will automatically indent for you when you hit the **RETURN** key.

Saving Files: Users can download Python notebooks onto their computer (notebook filenames end in .ipynb), they can also save files to the cloud or export straight to GitHub.

Writing Web Pages: Opening a Jupyter notebook enables the reader to download the file as a web page (web pages end in .html). You can then publish the notebook on the web through GitHub. Readers can easily find instructions online.

EXERCISES

1. Lists and dictionaries:

    (a) Given the list, A=[[1,2,3,4],[5,6,7,8],[9,10,11,12],[13,14,15,16]], how would you access the third element in the second list?

    (b) Consider A as a 4 × 4 array, slice A to remove row 3 and columns 1 and 2.

    (c) Using the **range** function, construct a list of the form $[-9, -5, -1, \ldots, 195, 199]$.

    (d) Create a dictionary data type for a car with key:value pairs, brand:BMW, year:2018, color:red, mileage:30000 and fuel:petrol.

2. Write a function for converting Kelvin to degrees Centigrade, giving the answer to four significant figures.

3. Defining mathematical functions using **def**:

    (a) Define the sigmoid function whose formula is given by: $\sigma(x) = \frac{1}{1+e^{-x}}$. Determine $\sigma(0.5)$. This is an activation function used in artificial intelligence.

    (b) Define the hsgn function given as:

$$\text{hsgn}(x) = \begin{cases} 1 & \text{if } x > 0 \\ 0 & \text{if } x = 0 \\ -1 & \text{if } x < 0. \end{cases}$$

    Write a Python program that defines this function and determine $\text{hsgn}(-6)$.

4. Longer Python programs:

    (a) Write an interactive Python program to play a "guess the number" game. The computer should generate a random integer between 1 and 20, and the user (player) has to try to guess the number within six attempts. The program should let the user know if the guess is too high or too low. Readers will need the randint function from the random library.

(b) Consider Pythagorean triples, positive integers a, b, c, such that $a^2 + b^2 = c^2$. Suppose that c is defined by $c = b + n$, where n is also an integer. Write a Python program that will find all such triples for a given value of n, where both a and b are less than or equal to a maximum value, m, say. For the case, $n = 1$, find all triples with $1 \leq a \leq 100$ and $1 \leq b \leq 100$. For the case, $n = 3$, find all triples with $1 \leq a \leq 200$ and $1 \leq b \leq 200$.

## CHAPTER 3

The Turtle Library

Run all of the programs in this chapter in Google Colab.

```
In [1]: # To insert an image in Google Colab.
# Turtle.png should be loaded in the Colab folder.
from google.colab import files
from IPython.display import Image
Image("Turtle.png", width = 600)
```

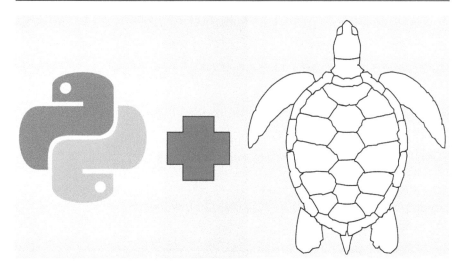

The reader can also run all of the programs in this chapter in Python IDLE, however, some of the syntax will be slightly different. The Python IDLE programs are listed in my other CRC Press book: Python for Scientific Computing and Artificial Intelligence, referenced in the Preface.

DOI: 10.1201/9781003472759-3

3.1   THE CANTOR SET FRACTAL

Start with a line segment at stage zero. At each stage, remove the middle third segment and replace one segment with two segments each one third the length of the previous segment.

```
In [2]: # You must run this cell before the other programs.
# These commands are not needed in IDLE.
!pip install ColabTurtlePlus
from ColabTurtlePlus.Turtle import *
```

```
In [3]: # The Cantor set.
initializeTurtle()
def cantor(x , y , length):
  speed(13)                    # Fastest speed.
  if length >= 5:
    penup()
    pensize(2)
    pencolor("blue")
    setpos(x , y)
    pendown()
    fd(length)                 # Forward.
    y -= 80                    # y = y - 80.
```

```
    cantor(x , y , length / 3)
    cantor(x + 2 * length / 3 , y , length / 3)
    penup()
    setpos(x , y + 80)
cantor(-400 , 200 , 500)
```

3.2   THE KOCH SNOWFLAKE

Start with an equilateral triangle. On the outside, replace each segment with four segments, each one-third the length of the previous segment.

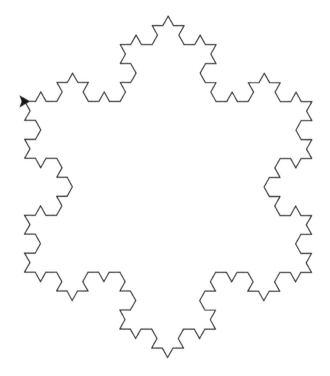

```
In [4]: # The Koch Snowflake.
initializeTurtle()
pensize(1)
penup()
setpos(-250 , 200)                   # To center image.
pendown()
def KochSnowflake(length, level):
    speed(13)                        # Fastest speed.
```

```
  for i in range(3):
    plot_side(length, level)
    rt(120)                              # Right turn 120 degrees.
def plot_side(length, level):
  if level==0:
    fd(length)
    return
  plot_side(length/3, level-1)
  lt(60)                                 # Left turn 60 degrees.
  plot_side(length/3, level-1)
  rt(120)
  plot_side(length/3, level-1)
  lt(60)
  plot_side(length/3, level-1)
KochSnowflake(300 , 3)
```

3.3 A BIFURCATING TREE

To plot a bifurcating tree, at each stage replace one segment with two
shorter segments, turning one left and the other right through given
angles.

```
In [5]: # A bifurcating tree.
initializeTurtle()
speed(13)
setheading(90)          # Point turtle up.
penup()                 # Lift pen.
setpos(-200 , -120)     # Start point coordinates.
pendown()
def FractalTreeColor(length, level):
  pensize(length / 10) # Thickness of lines.
  if length < 20:
    pencolor("green")
  else:
    pencolor("brown")
  if level > 0:
    fd(length)               # Forward
    rt(60)                   # Right turn 60 degrees
    FractalTreeColor(length*0.7, level-1)
    lt(90)                   # Left turn 90 degrees
```

```
    FractalTreeColor(length*0.5, level-1)
    rt(30)              # Right turn 30 degrees
    penup()
    bk(length)          # Backward
    pendown()
FractalTreeColor(200 , 10)
```

3.4   THE SIERPINSKI TRIANGLE

Start with a solid equilateral triangle. At each stage remove the inverted central equilateral triangle.

```
In [6]: # Sierpinski Triangle.
initializeTurtle()
speed(13)                    # Fastest speed.
penup()
setpos(-300 , 0)
pendown()
def SierpinskiTriangle(length, level):
  if level==0:
```

```
    return
  begin_fill()                # Fill shape.
  color("red")
  for i in range(3):
    SierpinskiTriangle(length/2, level-1)
    fd(length)
    lt(120)                   # Left turn 120 degrees.
  end_fill()
SierpinskiTriangle(300 , 5)
```

EXERCISES

1. Plot a variation of the Cantor set, where the two middle segments
   are removed at each stage. Thus, at each stage, one segment is replaced with three segments each on-fifth the length of the previous
   segment.

2. Plot a Koch square fractal, where on each side of a square, one segment is replaced with five segments each one-third the length of the previous segment.

3. Plot a trifurcating tree fractal.

4. Plot a Sierpinski square fractal, where at each stage a central square is removed and the length scales decrease by one-third.

5. In Python IDLE, click on the **Help** tab and **Turtle Demo** to see some cool examples. Figure 3.1 shows stills of three animations.

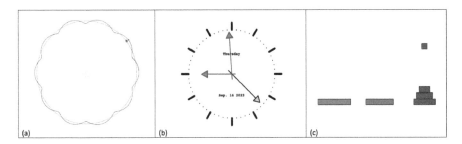

Figure 3.1 Stills of animations from the Turtle Demo. (a) Trajectories of the earth and moon around the sun. (b) Analog clock with day and date. (c) Solution to the Towers of Hanoi problem.

## CHAPTER 4

NumPy and MatPlotLib

4.1 NUMERICAL PYTHON (NUMPY)

Numpy is a library that allows Python to compute with lists, arrays, vectors, matrices and tensors. For more information, see the web pages at:

https://numpy.org/doc/stable/reference/.

Examples.

In/Out	Code Cells
In [1]:	```# Import numpy into the np namespace.``` ```import numpy as np```

In [2]:	```# Define an array.```
	```# No commas between the elements when printed.```
	```# A 5-dimensional vector and rank one tensor.```
	```A = np.arange(5))```

 DOI: 10.1201/9781003472759-4

```
Out[2]:   [0 1 2 3 4]
In [3]:   # Convert an array to a list.
          # Lists have commas between elements.
          A.tolist()
Out[3]:   [0, 1, 2, 3, 4]
In [4]:   # Two 2x2 arrays. These are rank two tensors.
          B = np.array([[1,1] , [0,1]])
          C = np.array([[2,0] , [3,4]])
          # Elementwise multiplication.
          B * C
Out[4]:   array([[2 , 0] , [0 , 4]])
In [5]:   # Matrix multiplication.
          np.dot(B , C)
Out[5]:   array([[5 , 4] , [3 , 4]])
In [6]:   # A 3x3 array and rank two tensor.
          D = np.arange(9).reshape(3 , 3)
          print(D) , D.ndim
Out[6:]   ([[0 1 2] [3 4 5] [6 7 8]] , 2)
In [7]:   # A rank three tensor.
          # A 2x2x2 array.
          T=np.array([[[1,2] , [3,4]], [[5,6],[7,8]]])
          T.ndim
Out[7]:   3
In [8]:   # Vectorized computation.
          D.sum(axis = 0) # Sum columns.
Out[8]:   array([9 , 12 , 15])
In [9]:   # The minimum of each row.
          D.min(axis = 1) # Work with rows.
Out[9]:   array([0 , 3 , 6])
In [10]:  # Slicing arrays.
          D[: , 1] # All elements in column 2.
Out[10]:  array[1 , 4 , 7]
In [11]:  # Slicing.
          E = np.arange(12).reshape(3 , 4)
          print(E)
Out[11]:  [[0 1 2 3] , [4,5,6,7] , [8,9,10,11]]
```

```
            # Make up your own examples for understanding.
In [12]:  E[:2 , 3:]
Out[12]:  array([[3] , [7]])
```

Vectorized Computation in NumPy: There is a single data type in NumPy arrays and this allows for vectorization. As far as the reader is concerned, for and while loops, which can use a lot of computer memory, can be replaced with vectorized forms. As a simple example, consider the list of numbers, $a = [1, 2, 3, \ldots, 100]$, and suppose we wish to compute $S_{100} = \sum_{n=1}^{100} a_i$. Then using for loops, the code would be:

```
In [13]: # Sum numbers using a for loop.
sum = 0
a= [i for i in range(101)]
for j in range(101):
    sum += a[j]
print(sum)
```

```
Out[13]: 5050
```

The vectorized version of this program is listed below:

```
In [14]: # Vectorized computation.
import numpy as np
a = np.arange(101)
print("The sum is" , np.sum(a))
```

```
Out[14]: The sum is 5050
```

4.2   MATPLOTLIB

MatPlotLib is an acronym for MATrix PLOTting LIBrary and it is a comprehensive library for creating animated, static, and more recently, interactive visualizations in Python. For more information the reader is directed to the URL:

```
https://matplitlib.org.
```

Example 4.2.1. Use MatPlotLib to plot the curve $y = x^2$, for $-2 \leq x \leq 2$. Save the figure as **Parabola.png** using 300 dots per square inch (dpi) resolution.

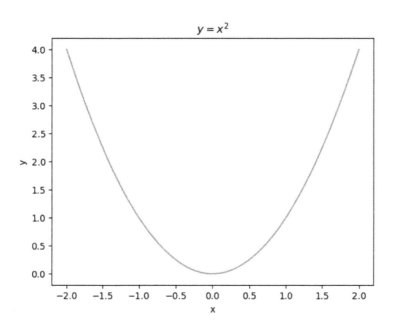

```
In [15]: # Simple plots in Python.
import numpy as np
import matplotlib.pyplot as plt
x = np.linspace(-2 , 2 , 100) # Define the domain values.
y = x**2                      # A vector of y-values.
plt.plot(x , y)
plt.xlabel("x")
plt.ylabel("y")
plt.title("$y=x^2$")
plt.savefig("Parabola.png" , dpi = 300)
plt.show()
```

Example 4.2.2. Use MatPlotLib to plot the curve $y = 4x^3 - 3x^4$, for $-2 \leq x \leq 2$ and $-5 \leq y \leq 2$.

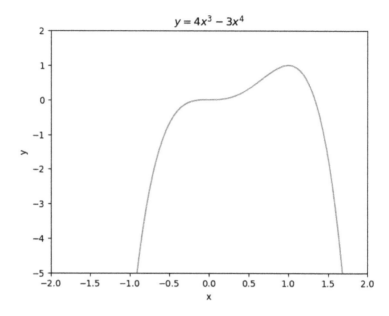

```
In [16]: # Plot the curve of a polynomial.
# Simple plots in Python.
import numpy as np
import matplotlib.pyplot as plt
x = np.linspace(-2 , 2 , 100) # Define the domain values.
y = 4 * x**3 - 3 * x**4        # A vector of y-values.
plt.axis([-2 , 2, -5 , 2])     # Set the ranges for x and y.
plt.plot(x , y)
plt.xlabel("x") , plt.ylabel("y")
plt.show()
```

4.3 SCATTER PLOTS

Scatter plots are used a lot in Data Science.

Example 4.3.1. Generate 50 random dots with random radii and colors.

```
In [17]: # Scatter plot.
import numpy as np
import matplotlib.pyplot as plt
# Fix the random state.
```

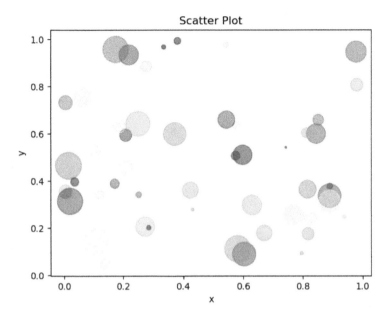

```
# You can change this to get a different plot.
np.random.seed(100)
N = 50
x , y = np.random.rand(N) ,  np.random.rand(N)
colors = np.random.rand(N)
area = (30 * np.random.rand(N))**2  # Change radii.
plt.scatter(x , y , s = area , c = colors , alpha=0.5)
plt.xlabel("x") , plt.ylabel("y")
plt.show()
```

4.4 SURFACE PLOTS

Example 4.4.1. Using the command **%matplotlib qt5**, opens an interactive window where readers can rotate the surface in 3D. In Spyder, one types **%matplotlib qt5** in the Command Window before plotting. Plot the surface, $z = x^2 + y^2$.

```
In [18]: # Plotting a surface in 3D.
# An interactive window will appear in this case.
%matplotlib qt5
from mpl_toolkits.mplot3d import axes3d
```

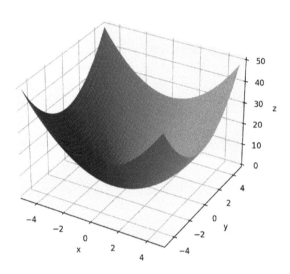

3D Surface, $z = x^2 + y^2$

```
import matplotlib.pyplot as plt
import numpy as np
x = np.arange(-5 , 5 , 0.1)
y = np.arange(-5 , 5 , 0.1)
X,Y = np.meshgrid(x,y)
# Z is a function of two variables.
Z = X**2 + Y**2
fig = plt.figure(figsize=(12,6))
ax = fig.add_subplot(111, projection = "3d")
ax.set_xlabel("x")
ax.set_xlim(-5 , 5)
ax.set_ylabel("y")
ax.set_ylim(-5 , 5)
ax.set_zlabel("z")
ax.set_zlim(np.min(Z) , np.max(Z))
ax.set_title("3D Surface, $z=x^2+y^2$")
ax.plot_surface(X , Y , Z)
plt.show()
```

EXERCISES

1. Plot the graph of the quadratic function: $y = f(x) = x^2 - 3x - 18$.

2. Plot the graph of the trigonometric function: $y = g(x) = 2 + 3\sin(x)$.

3. Plot the graph of the exponential function: $y = h(x) = 1 + e^{-x}$.

4. Plot the surface: $z = xe^{-(x^2+y^2)}$.

## CHAPTER 5

Google Colab, SymPy and GitHub

5.1 GOOGLE COLAB

Google Colab (Collaboratory) allows the user to write and execute Python code in their browser, with zero configuration required, free access to Graphical Processing Units (GPUs) and Tensor Processing Units (TPUs), and easy file sharing via GitHub. The user must have a Google account to use Colab, you can create a Google account here:

https://support.google.com/accounts/.

To use Google Colab, click here:

https://colab.research.google.com.

 DOI: 10.1201/9781003472759-5

The Google Colab interface is shown in the Preface. The major advantage with Colab is that users do not need any software stored on their computer and it can be accessed anywhere that has access to the internet.

5.2 FORMATTING NOTEBOOKS

In this section, the reader is shown how to insert titles, subtitles, bold font, italics, bullets and numbered items in a list in a Jupyter notebook.

The Text commands are shown in the left-hand window below and the corresponding output is shown on the right.

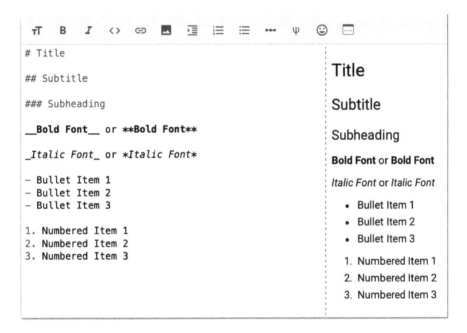

Mathematical Symbols are inserted into notebooks using LaTeX, a document preparation system used for producing highly professional documents. The mathematical symbols can also be inserted into figures as shown in other chapters of this book. Table 5.1 shows some popular LaTeX symbols for Jupyter notebooks and more symbols can be found here:

https://oeis.org/wiki/List_of_LaTeX_mathematical_symbols.

Table 5.1 Table of popular LaTeX symbols for Jupyter notebooks.

Greek Letters	LaTeX	Mathematics	LaTeX
α	\alpha	$\frac{dx}{dt}$	\frac{dx}{dt}
β	\beta	$\dot{x}$	\dot{ x }
γ, Γ	\gamma, \Gamma	$\ddot{x}$	\ddot{x}
δ	\delta	$\sin(x)$	\sin(x)
ϵ	\epsilon	$\cos(x)$	\cos(x)
θ	\theta	$\leq$	\leq
λ, Λ	\lambda, \Lambda	$\geq$	\geq
μ	\mu	x^2	x^2
σ, Σ	\sigma \Sigma	$\in$	\in
τ	\tau	$\pm$	\pm
ϕ	\phi	$\rightarrow$	\rightarrow
ω, Ω	\omega, \Omega	$\int$	\int

Some examples are listed below in a notebook. Note that the LaTeX commands are between dollar signs.

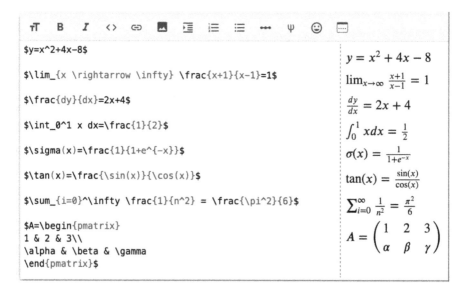

For more information on LaTeX, the reader is directed to Overleaf:

https://www.overleaf.com/,

the easy to use, and free, online collaborative LaTeX editor.

Inserting Figures: The commands used for inserting figures in Google Colab were given in Chapter 3. The figure below shows how to insert a figure in a Jupyter notebook. Note that the figure file "Python Logo.png" must be in the same folder where the Jupyter notebook is saved.

```
# HTML Code

The file "Python Logo.png" must be in the same folder
where the Jupyter notebook (.ipynb) is saved.|

<figure>
<img src = "Python Logo.png" , width = "600">
</figure>
```

5.3   SYMBOLIC PYTHON (SYMPY)

SymPy is a computer algebra system and a Python library for symbolic mathematics written entirely in Python. For more information, see the sympy help pages at:

https://docs.sympy.org/latest/index.html.

The examples below are taken from algebra, calculus and matrices. These topics are also covered in the next chapter. For more in-depth explanations, the reader is advised to use the web. Using Python can give a deeper understanding of the mathematics.

Examples.   1. Factorize $x^2 - y^2$.

2. Solve for x: $x^2 - 4x - 3 = 0$.

3. Solve the linear simultaneous equations: $x - y = 0, x + 2y - 5 = 0$.

4. Split into partial fractions: $\frac{1}{(x+1)(x+2)}$.

5. Find $\lim_{x \to 0} \frac{x}{\sin(x)}$.

6. Differentiate: $x^2 - 7x + 8$, with respect to x.

7. Determine the indefinite integral: $I = \int x^4 dx$.

8. Determine the definite integral: $I = \int_{x=1}^{3} x^4 dx$.

9. Compute π to 10 decimal places.

10. Given the matrices, $A = \begin{pmatrix} 1 & -1 \\ 2 & 3 \end{pmatrix}$ and $B = \begin{pmatrix} 0 & 2 \\ 3 & 3 \end{pmatrix}$, compute:

   (a) $2A + 3B$.
   (b) $A \times B$.
   (c) The inverse of matrix A, if it exists.
   (d) The determinant of A.

Solutions.

In/Out   Code Cells

In [1]:   ```
# Import all functions from the SymPy library.
from sympy import *
```

In [2]:    ```
# Declare symbolic objects.
x , y = symbols("x y")
```

In [3]:   ```
# 1. Factorization.
factor(x**2 - y**2)
```

Out[3]:    `(x - y)(x + y)`

In [4]:    ```
# 2. Solve a quadratic equation symbolically.
solve(x**2 - 4 * x - 3 , x)
```

Out[4]:   `[2 - sqrt(7) , 2 + sqrt(7)]`

In [5]:   ```
# 3. Solve simultaneous equations.
solve([x - y, x + 2 * y - 5])
```

Out[5]:    `{x: 5/3 , y: 5/3}`

In [6]:    ```
# 4. Partial fractions.
apart(1 /((x + 1 * (x + 2)))
```

Out[6:]   `-1/(x+2)+1/(x+1)`

In [7]:   ```
# 5. Limits, the limit as x goes to 0.
limit(x / sin(x) , x , 0)
```
```

```
Out[7]:    1
In [8]:    # 6. Differentiation (Calculus).
           diff(x**2 - 7 * x + 8 , x)
Out[8]:    2x - 7
In [9]:    # 7. Indefinite integration.
           print(integrate(x**4) , "+ c")
Out[9]:    x**5/5 + c
In [10]:   # 8. Definite integration.
           integrate(x**4 , (x , 1 , 3))
Out[10]:   242/5
In [11]:   # 9. Pi to 10 decimal places.
           N(pi , 10)
Out[11]:   3.1415926536
In [12]:   # 10. Define 2x2 matrices.
           A=Matrix([[1,-1] , [2,3]])
           B=Matrix([[0,2] , [3,3]])
           2 * A + 3 * B
Out[12]:   Matrix([[2,4] , [13,15]])
In [13]:   A * B # Multiplication.
Out[13]:   Matrix([[-3,-1] , [9,13]])
In [14]:   A.inv() # Inverse.
Out[14]:   Matrix([[3/5,1/5] , [-2/5,1/5]])
In [15]:   A.det() # Determinant.
Out[15]:   5
```

5.4 GITHUB

Readers are recommended to join GitHub:

https://github.com.

GitHub, Inc. is a platform and cloud-based service for programming development and version control, allowing coders to store and manage their code. As of January 2023, GitHub reported to have over one hundred million developers worldwide. It is used by many industries and universities.

The figure below shows the GitHub interface. There are plenty of videos on the web to get you started. You can save your Python files and notebooks here and even create your own web pages to publish.

Dr Stephen Lynch NTF
FIMA SFHEA
DrStephenLynch

Author of Python, Maple, MATLAB and Mathematica books. Inventor of Binary Oscillator Computing. Highly interdisciplinary research.

Edit profile

A 5 followers · 0 following

MMU
Manchester, UK
s.lynch@mmu.ac.uk
https://www.mmu.ac.uk/computing-and-maths/staff/profile/index.php?id=2443
@DrStephenLynch

EXERCISES

1. Use SymPy to:

   (a) Factorize $x^3 - y^3$.

   (b) Solve for x: $x^2 - 7x - 30 = 0$.

   (c) Split into partial fractions: $\frac{3x}{(x-1)(x+2)(x-5)}$.

   (d) Expand $(y + x - 3)(x^2 - y + 4)$.

   (e) Solve the linear simultaneous equations: $y = 2x + 2, y = -3x + 1$.

2. Compute the following:

   (a) $\lim_{x \to 1} \frac{x-1}{(x^2-1)}$.

   (b) The derivative of $y = x^2 - 6x + 9$.

   (c) The derivative of $y = \cos(3x)$.

   (d) The derivative of $y = 2e^x - 1$.

   (e) The derivative of $y = x \sin(x)$.

3. Determine the following integrals:

(a) $\int x^5 dx$.

(b) $\int_{x=1}^{4} x^5 dx$.

(c) $\int \cos(3x) dx$.

(d) $\int_{x=0}^{1} x \sin(x) dx$.

(e) $\int_{x=1}^{\infty} \frac{1}{x} dx$.

4. Given the matrices, $A = \begin{pmatrix} 1 & 1 \\ -1 & 0 \end{pmatrix}$ and $B = \begin{pmatrix} 1 & -3 \\ 4 & 7 \end{pmatrix}$, compute:

(a) $2A$.

(b) $3A + 4B$.

(c) $A \times B$.

(d) The inverse of matrix A, if it exists.

(e) The determinant of B.

5. Sign up to GitHub and upload your notebooks and Python files to your repository.

Python for Mathematics

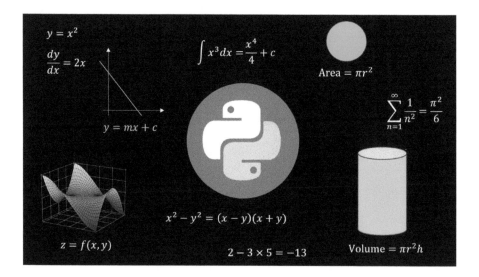

6.1 BASIC ALGEBRA

Algebra is a branch of mathematics in which abstract symbols, rather than numbers, are manipulated or operated with arithmetic. Children as young as ten years of age, or even younger, are generally introduced to algebra. Using Python can give a deeper understanding of the mathematics.

Example 6.1.1. Simple algebra.

1. Evaluate $y = mx + c$, given that $m = 2, x = 2, c = -1$.

DOI: 10.1201/9781003472759-6

2. Expand $2x(x-5)$.

3. Expand $(x-1)(x+3)$.

4. Factorize $2x^3 - 4x^2$.

5. Factorize $x^2 - x - 12$.

Solutions.

In/Out	Code Cells

In [1]:	`# 1. Substitution.`
	`m , x , c = 2 , 2 , -1`
	`y = m * x + c`

Out[1]:	`y = 3`

In [2]:	`# 2. Import symbols, expand and factor functions.`
	`from sympy import symbols , expand , factor`
	`# Expansion. Declare x to be symbolic.`
	`x = symbols("x")`
	`expand(2 * x * (x - 5))`

Out[2]:	`2 * x**2 - 10 * x`
In [3]:	`# 3. Expansion.` `expand((x - 1) * (x + 3))`
Out[3]:	`x**2 + 2 * x - 3`
In [4]:	`# 4. Factorization.` `factor(2 * x**3 - 4 * x**2)`
Out[4]:	`2 * x**2 * (x - 2)`
In [5]:	`# Factorize a quadratic.` `factor(x**2 - x - 12)`
Out[5]:	`(x - 4) * (x + 3)`

6.2 SOLVING EQUATIONS

The **solve** command in Python can solve linear and some nonlinear simultaneous equations as illustrated below:

Example 6.2.1. Solving equations.

1. Solve for x, $\frac{x+1}{x-1} = 2$.

2. Solve for y, $\frac{y-2}{3} = -\frac{4y-7}{6}$.

3. Plot the lines $\ell_1 : 2y = 4x + 5$ and $\ell_2 : 4y = 3x + 5$, on one graph and determine where they intercept.

4. Solve the quadratic equation: $2x^2 - 3x - 2 = 0$.

5. Determine where the line $y = 2x - 5$, meets the parabola, $y = x^2 - 4x + 3$.

Solutions.

In/Out   Code Cells

```
In [1]:   # 1. Solve for x.
          x , y = symbols("x y")
          solve((x + 1) / (x - 1) - 2 , x)
```
Out[1]:   [3]
```
In [2]:   # 2. Solve for y.
          solve((y - 2) / 3 + (4 * y - 7) / 6 , y)
```
Out[2]:   [11/6]

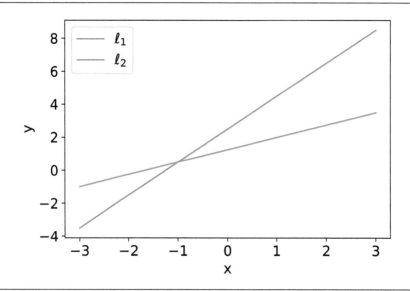

```
# In [3]: Plotting lines and finding where they cross
(see figure).
```

```
import numpy as np
from sympy import *
import matplotlib.pyplot as plt
x = np.linspace(-3 , 3 , 100) # Vector of x values.
y1 = (4 * x + 5) / 2
y2 = (3 * x + 5) / 4
plt.plot(x , y1 , label = "$\ell_1$")  # Curly letter 1.
plt.plot(x , y2 , label = "$\ell_2$")
plt.rcParams["font.size"] = 14 # Set font sizes.
plt.xlabel("x")
plt.ylabel("y")
plt.legend()
plt.show()
```

In/Out   Code Cells

In [4]:   # 3. Solving simultaneous equations.
          solve([2*y-4*x-5 , 4*y-3*x-5] , [x , y])
Out[4]:   {x: -1, y: 1/2}
In [5]:   # 4. Solve a quadratic equation.
          solve(2 * x**2 - 3 * x - 2)
Out[5]:   [-1/2 , 2]

```
# In [6]: Where a line meets a parabola.
sol = solve(x**2 - 4 * x + 3 - (2 * x - 5) , x)
print("The curves meet at coordinates: " , \
"(" , sol[0] , "," , 2 * sol[0] - 5 , ") and " \
 "(" , sol[1] , "," , 2 * sol[1] - 5 , ")" )
```

Out[6]: The curves meet at coordinates: $(2 , -1)$ and $(4 , 3)$

6.3 FUNCTIONS (MATHEMATICS)

A function is a rule that assigns to every x-value in the domain, one and only one y-value in the range.

Example 6.3.1. Use numpy and matplotlib to plot:

1. A graph of the function, $f(x) = x^2 - 4x - 1$.

2. Determine the x-values where $f(x) = 1$.

3. Determine the inverse function, $g^{-1}(x)$, of the function $y = g(x) = \frac{2x-1}{x+1}$. Plot $g(x)$ and $g^{-1}(x)$ on the same graph.

4. Given that $\alpha(x) = 2x - 5$ and $\beta(x) = 1 - x^2$, determine the composite functions, $\alpha(\beta(x))$ and $\beta(\alpha(x))$.

```
# In [1]: Define a function and plot its curve.
import matplotlib.pyplot as plt
import numpy as np
def f(x):
    return x**2 - 4 * x - 1
x = np.linspace(-2 , 6 , 100)
y = x**2 - 4 * x - 1
plt.rcParams["font.size"] = 14
plt.xlabel("x") , plt.ylabel("f(x)")
plt.plot(x , y)
plt.show()
```

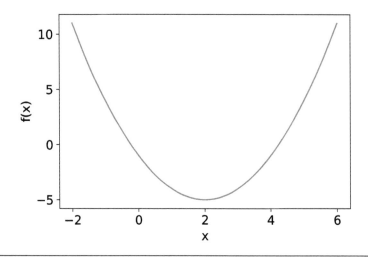

```
# In [2]: Determine where f(x)=1.
from sympy import *
x = symbols("x")
sol = solve(x**2 - 4 * x - 1 - 1 , x)
print("y=f(x)=1, when ", "x =" , sol[0], "and x =" , sol[1])
```

Out[2]: y=f(x)=1, when x = 2 - sqrt(6) and x = 2 + sqrt(6)

```
# In [3]: Determine an inverse function.
from sympy import symbols
x , y = symbols("x y")
sol = solve((2 * y -1) / (y + 1) - x , y)
print("Inverse function is: y=" , sol)
```

Out[3]: Inverse function is: y= [(-x - 1)/(x - 2)]

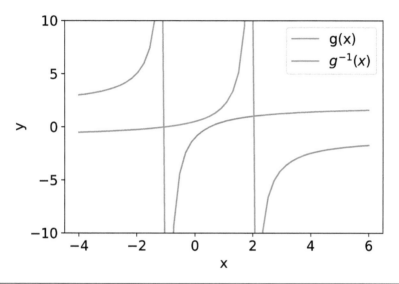

```
# In [4]: Plot of g(x) and its inverse function.
# Note here that vertical asymptotes are plotted.
import matplotlib.pyplot as plt
import numpy as np
x = np.linspace(-4 , 6 , 50)
g = (2 * x - 1) / (x + 1)
ginv = (-x - 1) / (x - 2)
plt.rcParams["font.size"] = 14
plt.ylim(-10 , 10)
plt.xlabel("x")
plt.ylabel("y")
plt.plot(x , g , label = "g(x)")
plt.plot(x , ginv , label = "$g^{-1}(x)$")
plt.legend()
plt.show()
```

```
# In [5]: Composition of functions (functions of functions).
from sympy import symbols
x = symbols("x")
def alpha(x):
    return 2 * x - 5
def beta(x):
    return 1 - x**2
print("alpha(beta(x) = " , alpha(beta(x)))
print("beta(alpha(x) = " , beta(alpha(x)))
```

Out[5]: alpha(beta(x)) = -2*x**2 - 3, beta(alpha(x)) = 1 - (2*x - 5)**2

6.4 DIFFERENTIATION AND INTEGRATION (CALCULUS)

Differentiation is used at this level to determine rates of change, or gradients of tangents to curves.

Integration is the opposite of differentiation and can be used to determine the area under curves.

Both topics come under the umbrella of calculus.

Example 6.4.1. Calculus:

1. Differentiate, or find $\frac{dy}{dx}$, the derivative of, $y = x^2 + 4x - 3$.

2. Determine the gradient of the tangent to the curve, $y = 4x^3 - 3x^4$, at the point $x = -\frac{1}{2}$. Plot the curve and its tangent.

3. Integrate, $I = \int 2x + 5\,dx$.

4. Determine the area under the curve, $y = 4x^3 - 3x^4$, between $x = 0.5$ and $x = 1$ and above the x-axis. Plot the curve and shade the area.

In/Out   Code Cells

```
In [1]:    # 1. Differentiation.
           from sympy import *
           x = symbols("x")
           diff(x**2 + 4 * x - 3 , x)
Out[1]:    2x + 4
```

```
# In [2]: 2. Find the gradient at x=-1/2.
# Use the subs function from sympy.
x , y , c = symbols("x y c")
xval = -1 / 2
y = 4 * x**3 - 3 * x**4
dydx = diff(y , x)
# The gradient (m) of the tangent.
m = dydx.subs(x , xval)
print("m = " , m)
yval = y.subs(x , xval)
print("y-value = " , yval)
# Determine the y-intercept (c) for the tangent line.
c = yval - m * xval
print("c = " , c)
# Plot the curve and tangent line.
x = np.linspace(-1 , 1.5 , 100)
y = 4 * x**3 - 3 * x**4
plt.rcParams["font.size"] = 14
plt.plot(x , y , label = "$y=4x^3-3x^4$")
plt.plot(x , m * x + c , label = "y=mx+c, tangent")
plt.legend()
plt.xlabel("x")
plt.ylabel("y")
plt.show()
```

```
Out[2]: m =  4.5000, y-value =  -0.6875, c =  1.5625
```

In/Out   Code Cells

```
In [3]:    # 3. Indefinite integration.
           x = symbols("x")
           print("I = " , integrate(2 * x + 5 , x) , "+ c")
Out[3]:    I = x**2 + 5 * x + c
```

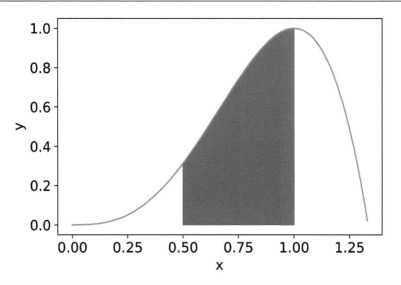

```
# In [4]: 4. Integration and finding the area under a curve.
A = integrate(4 * x**3 - 3 * x**4 , (x , 0.5 , 1))
print("Area = " , A , "units squared")
# Plot a figure.
import numpy as np
from matplotlib import pyplot as plt
def f(x):
    return 4 * x**3 - 3 * x**4
x = np.arange(0 , 1.34 , 0.01)
plt.rcParams["font.size"] = 14
plt.plot(x , f(x))
# Shade the area.
plt.fill_between(
        x = x,
```

```
        y1 = f(x),
        where = (0.5 <= x) & (x <= 1), # Limits of
                                           integration.
        color = "b")                    # Color of shading.
plt.xlabel("x")
plt.ylabel("y")
plt.show()
```

Out[4]: Area = 0.35625 units squared

Python for A-Level Mathematics and Beyond: I have published a Jupyter notebook online that shows how Python covers the whole A-Level Mathematics syllabus in the UK:

https://drstephenlynch.github.io/webpages/Python_for_A_Lev
el_Mathematics_and_Beyond.html.

EXERCISES

1. Basic algebra:

   (a) Given that $s = ut + \frac{1}{2}at^2$, determine s if $u = 1, t = 1, a = 9.8$.

   (b) Expand $2x(x - 4)$.

   (c) Expand $(x + y + 1)(x - y + 2)$

   (d) Factorize $x^2 - 7x - 12$.

2. Solving equations:

   (a) Solve for t given, $3 = 5 + 10t$.

   (b) Solve for a given, $s = ut + \frac{1}{2}at^2$.

   (c) Solve the quadratic equation: $2x^2 + x - 3 = 0$.

   (d) Determine where the line, $y = x$, meets the circle $x^2 + y^2 = 1$.

3. Functions:

   (a) Given $f(x) = \frac{2x+3}{x-5}$, determine $f(4)$.

   (b) Plot the function, $f(x) = \frac{2x+3}{x-5}$, and determine where $f(x) = 1$.

(c) Given that, $g(x) = 3x + 4$ and $h(x) = 1 - x^2$, find $g(h(x)) - h(g(x))$.

(d) The logistic map function is given by, $f_\mu(x) = \mu x(1 - x)$, compute $f_\mu(f_\mu(x))$.

4. Calculus:

(a) If $y = \sin(2x)$, compute $\frac{dy}{dx}$.

(b) Determine $\frac{dy}{dx}\big|_{x=0}$ for $y = x^3 - 1$.

(c) Determine the area bound by the curves $y = 1 - x^2$ and $y = x^2 - 1$.

(d) Plot the curves for the previous example and shade the area.

Introduction to Cryptography

Cryptography (cyber security) is a method of protecting information and communications through the use of codes, so that only those for whom the information is intended can read and process it.

7.1 THE CAESAR CIPHER

Caesar's cipher is one of the simplest encryption techniques and is often used by children when learning to send secret messages. It is a type of substitution cipher in which each letter in the plaintext is replaced by a letter some fixed number of positions down the alphabet.

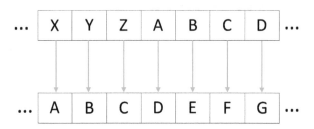

https://en.wikipedia.org/wiki/Caesar_cipher.

Unicode is a universal character encoding standard:

https://en.wikipedia.org/wiki/List_of_Unicode_characters.

The ord() function in Python takes a string argument of a single Unicode character and returns its integer. The chr() function is the inverse of the ord() function. For example, the ord("A") is 65, and the ord("a") is 97.

```python
#  In [1]: The Caesar Cipher: Encryption.
shift = 3
def encrypt(text , shift):
    result = ""
    for i in range(len(text)):
        char = text[i]
        if (char.isupper()): # Upper-case characters.
            result += chr((ord(char) + shift - 65) % 26 + 65)
        else:                # Lower-case characters.
            result += chr((ord(char) + shift - 97) % 26 + 97)
    return result
# Insert text.
text = "Caeser Cipher Code"
print("Text : " + text)
```

```
print("Shift : " + str(shift))
print("Cipher: " + encrypt(text,shift))
```

In/Out   Code Cells

```
Out[1]:   Text : Caesar Cipher Code
          Shift : 3
          Cipher : FdhvhuqFlskhuqFrgh
```

7.2   THE XOR CIPHER

The table below is the truth table for the Exclusive OR (XOR) function. In Python, the operator used is X^Y, read as X XOR Y.

Bitwise XOR (^)

X	Y	X^Y
0	0	0
0	1	1
1	0	1
1	1	0

Example 7.2.1. Determine 3 ^ 6.

Solution. Now in binary, $3 = 0011$ and $6 = 0110$. Use bitwise XOR on the columns:

```
0  0  0  1  1
0  0  1  1  0
───────────────
0  0  1  0  1
```

Thus, 0011 ^ $0110 = 0101$, or 3 ^ $6 = 5$.

In/Out   Code Cells

```
In [2]:   3 ^ 6 , 0b011 ^ 0b110
Out[2]:   (5 , 5)
```

Example 7.2.2. Use Python to encrypt and decrypt using the XOR cipher.

Solution. The program is listed below:

```
# In [3]: XOR Cipher - Encryption and Decryption.
# The same function is used to encrypt and decrypt.
def EncryptDecrypt(text):
    XORKey = "A"
    length = len(text)
# Perform XOR operation of key with every character.
    for i in range(length):
        text = (text[:i] + chr(ord(text[i]) ^ ord(XORKey)) \
                + text[i + 1:])
        print(text[i], end = "")
    return text
if __name__ == "__main__":
    sampletext = "The Exclusive Or Cipher"
    print("Encrypted String: ", end = "")
    sampletext = EncryptDecrypt(sampletext) # Encrypt.
    print("\n")
    print("Decrypted String: ", end = "")
    EncryptDecrypt(sampletext)                  # Decrypt.
```

```
Encrypted String: □)$a□9"-42(7$a□3a□(1)$3

Decrypted String: The Exclusive Or Cipher
```

7.3 THE RIVEST-SHAMIR-ADLEMAN (RSA) CRYPTOSYSTEM

Basically, RSA relies on the problem of factoring the product of two large prime numbers (in industry, each prime number typically has 10^{300} digits), and breaking RSA encryption is known as the RSA problem. For more details see Wikipedia:

https://en.wikipedia.org/wiki/RSA_(cryptosystem).

Definition 7.3.1: A **prime** number is a number that can only be divided by itself and one.

Definition 7.3.2: Given an integer $k > 1$, called a **modulus**, two integers m and n, are said to be **congruent** modulo k, if k is a divisor of their difference, so $m - n = rk$, where r is an integer.

Definition 7.3.3: Congruence modulo k is written: $m \equiv n \bmod(k)$.

Definition 7.3.4: The **Extended Euclidean Algorithm (EEA)** computes, in addition to the greatest common divisor (gcd) of integers a and b, also the coefficients of Bézout's identity, which are integers x and y such that:

$$ax + by = \gcd(a, b).$$

Readers can find more information on Wikipedia:

https://en.wikipedia.org/wiki/Extended_Euclidean_algorithm.

The RSA algorithm can be described in five simple steps:

1. Choose two large prime numbers, p and q, say.

2. Compute $n = p \times q$.

3. The Euler totient function is given as: $\phi(n) = (p - 1) \times (q - 1)$.

4. The public key, e say, is chosen such that $2 < e < \phi(n)$, and $\gcd(e, \phi(n)) = 1$; that is, e and $\phi(n)$ are co-prime.

5. The public and private keys satisfy the congruence, $d \equiv e^{-1} \bmod(\phi(n))$, where d is the private key computed using the EEA.

7.4 SIMPLE RSA ALGORITHM EXAMPLE

The RSA algorithm for encryption and decryption can be summarized using the following steps:

1. **Bob** follows the steps 1 to 5 in the RSA algorithm above.

2. **Alice** computes $E = m^e \bmod(n)$, where m is the message and E is the encrypted message, and sends E to Bob.

3. **Bob** computes $D = E^d \bmod(n)$, where D is the decrypted message.

Historically, Alice and Bob were the names given to fictitious characters used to explain how the RSA encryption method worked.

Example 7.4.1. Two small prime numbers are chosen for illustrative purposes. In real-world applications, very large prime numbers are chosen. Bob chooses the prime numbers, $p = 17$ and $q = 29$, and a public key, $e = 11$. Alice wishes to send the number (message) 20 to Bob. What is the encrypted message sent? Confirm that Bob correctly decrypts the message.

In/Out   Code Cells

```
In [4]:    # Check that p and q are prime.
           from sympy import *
           isprime(17) , isprime(29)
Out[4]:    (True , True)
```

```
# In [5]: RSA Algorithm Example.
from math import gcd
# Step 1, choose two prime numbers.
p , q = 17 , 29
# Step 2, compute n.
n = p * q
print("n =", n)
# Step 3, compute phi(n).
phi = (p - 1) * (q - 1)
# Step 4, compute a public key, e say,
# there are many possibilities.
# In this case, choose e = 11.
#for e in range(2 , phi):
#    if (gcd(e, phi) == 1):
#        print("e =", e)
e = 11
# Step 5:
# Python program for the EEA.
def extended_gcd(e, phi):
    if e == 0:
        return phi, 0, 1
```

```
    else:
        gcd, x, y = extended_gcd(phi % e, e)
        return gcd, y - (phi // e) * x, x
if __name__ == '__main__':
    gcd, x, y = extended_gcd(e, phi)
    print("phi = " , phi)
    print('The gcd is', gcd)
    print("x = " , x , ", y = " , y)
```

Out[5]: n=493, phi = 448. The gcd is 1; x = 163, y = -4 .

In order to encrypt and decrypt, one has to compute $x^y \bmod(z)$. In Python, we use the built-in function **pow(x,y,z)**, which takes three arguments.

IMPORTANT: If you import the pow function from the math library it only takes two arguments, and computes, $\mathrm{pow}(x, y) = x^y$.

```
# In [6]: Encryption and decryption.
# The message here is m = msg = 20.
d , e , n = 163 , 11 , 493
msg = 20
print(f"Original message: {msg}")
# Encryption
E = pow(msg , e , n)
print(f"Encrypted message: {E}")
# Decryption
D = pow(E , d , n)
print(f"Decrypted message: {D}")
```

Out[6]: Original message: 20 ; Encrypted message: 7 ; Decrypted message: 20.

EXERCISES

1. Write a Python program to decrypt the Caesar cipher.

2. Research the Playfair Cipher on the Web:

`https://en.wikipedia.org/wiki/Playfair_cipher.`

Use your first name and surname for the encryption and decryption key (without repetition of letters) and complete a 5×5 array table, missing out the letter J. For example, if your name is **Bruce Wayne**, the array would look like:

$$\begin{pmatrix} B & R & U & C & E \\ W & A & Y & N & D \\ F & G & H & I & K \\ L & M & O & P & Q \\ S & T & V & X & Z \end{pmatrix}.$$

Write down the rules of the Playfair Cipher. Given that the message is "I LOVE PYTHON," determine the encrypted message by hand. Use the letter X for spaces.

3. Research the Vigenère Cipher on the Web:

`https://en.wikipedia.org/wiki/Vigenere_cipher.`

Write a Python program to encrypt a message using this cipher.

4. For those interested in cryptography, carry out a literature search on:

(a) Symmetric Cryptography: block ciphers, stream ciphers, hash functions, keyed hashing and authenticated encryption.

(b) Asymmetric Cryptography: Rivest-Shamir-Adleman (RSA) a public-key cryptosystem, Diffie-Hellman key exchange, authenticated encryption, elliptic curve cryptography and chaos synchronization cryptography.

(c) For this problem, use the RSA public key cryptosystem. Bob chooses the prime numbers $p = 19$, $q = 23$ and $e = 7$. Alice wishes to send the number 11 to Bob. What is the message sent? Confirm that Bob correctly decrypts this.

An Introduction to Artificial Intelligence

Artificial Intelligence (AI) refers to computational systems that behave in a similar manner to the human brain.

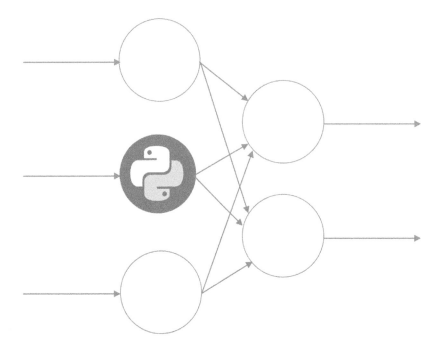

DOI: 10.1201/9781003472759-8

8.1 ARTIFICIAL NEURAL NETWORKS

Artificial Neural Networks (ANNs) are the building blocks for AI. The figure below shows a pictorial representation of a single mathematical neuron, where $x_1, x_2, \ldots, x_n$ are inputs, $w_1, w_2, \ldots, w_n$ represent weights (synaptic weights for biological neurons) b is a bias, v is an activation potential, defined by:

$$v = x_1 w_1 + x_2 w_2 + \ldots + x_n w_n + b,$$

$\sigma(v)$ is a transfer (or activation) function known as the sigmoid function:

$$\sigma(v) = \frac{1}{1 + e^{-v}},$$

and y is the output of the neuron, defined by:

$$y = \sigma(v).$$

Neurons are connected together to form neural networks.

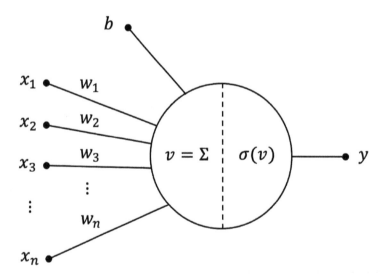

This single neuron ANN is also known as a perceptron. For difficult problems in AI, ANNs consist of many layers with many neurons in each layer, these are known as deep learning ANNs. AI is inspired by the dynamics of neurons in the human brain.

8.2 THE AND/OR AND XOR GATE ANNS

Figures 8.1 to 8.3 show the ANNs and corresponding truth tables for the AND, OR and XOR logic gates, respectively.

The AND Gate ANN A single neuron can model an AND gate ANN as illustrated below.

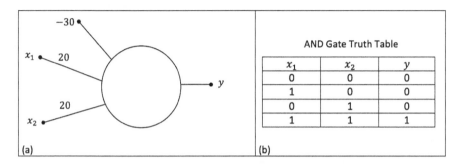

| | AND Gate Truth Table | |
x_1	x_2	y
0	0	0
1	0	0
0	1	0
1	1	1

(a) (b)

Figure 8.1 (a) ANN for an AND gate. (b) Truth table for an AND gate.

```
In [1]: # Python Program for the AND Gate ANN.
import numpy as np
w1 , w2  = 20 , 20
b = -30
# Define the functions.
def sigmoid(v):
    return 1 / (1 + np.exp(- v))
def AND(x1, x2):
    return sigmoid(x1 * w1 + x2 * w2 + b)
print("AND(0,0)=", AND(0,0))
print("AND(1,0)=", AND(1,0))
print("AND(0,1)=", AND(0,1))
print("AND(1,1)=", AND(1,1))
```

```
Out[1]: AND(0,0)= 9.357622968839299e-14
AND(1,0)= 4.5397868702434395e-05
AND(0,1)= 4.5397868702434395e-05
AND(1,1)= 0.9999546021312976
```

The OR Gate ANN A single neuron can also model an OR gate ANN.

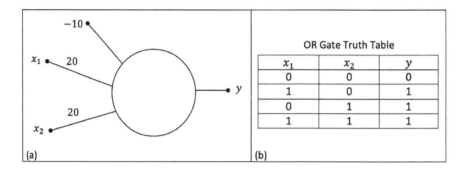

Figure 8.2 (a) ANN for an OR gate. (b) Truth table for an OR gate.

```
In [2]: # Python Program for the OR Gate ANN.
import numpy as np
w1 , w2 = 20 , 20
b = -10
# Define the functions.
def sigmoid(v):
    return 1 / (1 + np.exp(- v))
def OR(x1, x2):
    return sigmoid(x1 * w1 + x2 * w2 + b)
print("OR(0,0)=", OR(0,0))
print("OR(1,0)=", OR(1,0))
print("OR(0,1)=", OR(0,1))
print("OR(1,1)=", OR(1,1))
```

```
Out[2]: OR(0,0)= 4.5397868702434395e-05
OR(1,0)= 0.9999546021312976
OR(0,1)= 0.9999546021312976
OR(1,1)= 0.9999999999999065
```

The XOR Gate ANN This cannot be modeled using a single neuron, a hidden layer must be introduced as shown in Figure 8.3.

You will be given the weights and the biases and asked to write a Python program for an XOR gate ANN in the exercises at the end of the chapter.

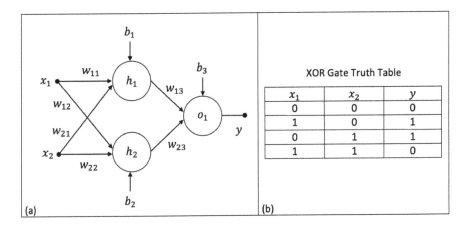

Figure 8.3 (a) ANN for an XOR gate. (b) Truth table for an XOR gate.

An AND gate and an XOR gate can be connected together to make a binary half-adder, used to add two bits together. This is the simplest logic component of all modern computers.

8.3 THE BACKPROPAGATION ALGORITHM

The figure below shows an ANN used to demonstrate how feedforward and backpropagation work in AI.

$$b_1 \qquad \sigma(h_1) = \frac{1}{1+e^{-h_1}} \qquad b_2 \qquad Err = \frac{1}{2}(y_t - y)^2$$

$$x_1 \xrightarrow{\quad w_1 \quad} h_1 \xrightarrow{\quad w_2 \quad} o_1 \longrightarrow y = \sigma(o_1)$$

$$h_1 = x_1 w_1 + b_1 \qquad o_1 = \sigma(h_1)w_2 + b_2$$

In the examples considered thus far, the weights for the ANNs have been given, but what if we are not given those weights? The backpropagation of errors algorithm is used to determine those weights. To illustrate the method, consider the above figure and use the transfer function $\sigma(v)$, furthermore suppose that the biases b_1 and b_2 are constant. Suppose we wish to update weights w_1 and w_2. Initially, random weights are chosen, data is fed forward through the network, and an error inevitably results. In this example, we use a mean squared error. Next, the backpropagation algorithm is used to update the weights in order to minimize the error. Thus, using the chain rule from calculus:

$$\frac{\partial Err}{\partial w_2} = \frac{\partial Err}{\partial y}\frac{\partial y}{\partial w_2} = \frac{\partial Err}{\partial y}\frac{\partial y}{\partial o_1}\frac{\partial o_1}{\partial w_2},$$

and

$$\frac{\partial Err}{\partial w_2} = (y_t - y) \times \frac{\partial \sigma}{\partial o_1} \times \sigma(h_1),$$

and

$$\frac{\partial Err}{\partial w_2} = (y_t - y) \times \sigma(o_1)(1 - \sigma(o_1)) \times \sigma(h_1).$$

The weight w_2 is then updated using the formula:

$$w_2 = w_2 - \eta \times \frac{\partial Err}{\partial w_2},$$

where η is called the learning rate. A similar argument is used to update w_1, thus:

$$\frac{\partial Err}{\partial w_1} = \frac{\partial Err}{\partial y}\frac{\partial y}{\partial w_1} = \frac{\partial Err}{\partial y} \times \frac{\partial y}{\partial o_1} \times \frac{\partial o_1}{\partial \sigma(h_1)} \times \frac{\partial \sigma(h_1)}{\partial h_1} \times \frac{\partial h_1}{\partial w_1},$$

and

$$\frac{\partial Err}{\partial w_1} = (y_t - y) \times \sigma(o_1)(1 - \sigma(o_1)) \times w_2 \times \sigma(h_1)(1 - \sigma(h_1)) \times x_1.$$

The weight w_1 is then updated using the formula:

$$w_1 = w_1 - \eta \times \frac{\partial Err}{\partial w_1}.$$

Readers should note that partial differentiation is used here as the error, Err, is a function of y, σ, o_i, h_i and w_i, where o_i, h_i are output and hidden activation potentials, respectively. The basic idea is to minimize Err with respect to the weights w_1 and w_2, in this case. Note that the biases b_1, b_2 are assumed to be constant for simplification.

Example 8.3.1. Consider the ANN for the XOR gate. Given that, $w_{11} = 0.2$, $w_{12} = 0.15$, $w_{21} = 0.25$, $w_{22} = 0.3$, $w_{13} = 0.15$, $w_{23} = 0.1$, $b_1 = b_2 = b_3 = -1$, $x_1 = 1$, $x_2 = 1$, $y_t = 0$ and $\eta = 0.1$:

(i) Determine the output y of the ANN on a forward pass given that the transfer function is $\sigma(v)$.

(ii) Use the backpropagation algorithm to update the weights w_{13} and w_{23}, assuming that the biases are constant.

Solution. The program below gives $y = 0.28729994077761756$, and new weights $w_{13} = 0.1521522763401024$ and $w_{23} = 0.10215227634010242$. The reader will be asked to update the other weights in the exercises at the end of the chapter.

```
In [3]: # Simple feedforward and backpropagation.
import numpy as np
w11,w12,w21,w22,w13,w23 = 0.2,0.15,0.25,0.3,0.15,0.1
b1 , b2 , b3 = -1 , -1 , -1
yt , eta = 0 , 0.1
x1 , x2 = 1 , 1
def sigmoid(v):
    return 1 / (1 + np.exp(- v))
h1 = x1 * w11 + x2 * w21 + b1
h2 = x1 * w12 + x2 * w22 + b2
o1 = sigmoid(h1) * w13 + sigmoid(h2) * w23 + b3
y = sigmoid(o1)
print("y = ", y)
# Backpropagate.
dErrdw13=(yt-y)*sigmoid(o1)*(1-sigmoid(o1))*sigmoid(h1)
w13 = w13 - eta * dErrdw13
print("w13 = ", w13)
dErrdw23=(yt-y)*sigmoid(o1)*(1-sigmoid(o1))*sigmoid(h2)
w23 = w23 - eta * dErrdw23
print("w23 = ", w23)
```

```
Out[3]: y =   0.28729994077761756
w13 =   0.1521522763401024
w23 =   0.10215227634010242
```

8.4   BOSTON HOUSING DATA

The data was originally published by Harrison, D. and Rubinfeld, D.L. (1978) "Hedonic prices and the demand for clean air," J. Environ. Economics & Management, 5 , 81-102. There are 14 attributes for 506 houses, which is a small data set. The attributes include per capita crime rate by town, nitrous oxide concentration (parts per 10 million), index of accessibility to radial highways, for example, and the target data is median value of owner-occupied homes in thousands of dollars. The Boston housing data and a full working program of a Boston housing ANN calculator can be downloaded through the author's GitHub site:

https://github.com/proflynch/A-Simple-Introduction-to-Python,

where the data file is labeled "housing.txt" and the Python program is called "Boston-Housing-ANN-Calculator.ipynb." Once trained, the ANN can be used to value new houses added to the data set. Please note that this data is from 1978, so is out-of-date.

Readers should note that data may not always be accurate, even when it has been used to test algorithms for many years! For example, the target data seems to be censored at 50.00 (corresponding to a median price of 50,000 dollars); censoring is suggested by the fact that the highest median price of exactly 50,000 dollars is reported in 16 cases, while 15 cases have prices between 40,000 dollars and 50,000 dollars, with prices rounded to the nearest hundred. Harrison and Rubinfeld do not mention any censoring.

Readers may be interested in the specialist Python AI deep learning frameworks, **TensorFlow** and **PyTorch**. TensorFlow is a very powerful and mature deep learning framework whereas PyTorch is relatively new with a stronger community movement and it is more Python-friendly. For more information on AI, machine learning, deep learning and TensorFlow, the reader is directed to my book "Python for Scientific Computing and Artificial Intelligence," referenced in the Preface.

EXERCISES

1. Given the activation functions:

$$\sigma(v) = \frac{1}{1 + e^{-v}}, \qquad \phi(v) = \tanh(v) = \frac{e^v - e^{-v}}{e^v + e^{-v}},$$

show that:

(a) $\frac{d\sigma}{dv} = \sigma(v)(1 - \sigma(v))$;

(b) $\frac{d\phi}{dv} = 1 - (\phi(v))^2$.

Plot the curves and their derivatives.

2. Show that the XOR gate ANN acts as a good approximation of an XOR gate, given that:

$$w_{11} = 60, w_{12} = 80, w_{21} = 60, w_{22} = 80, w_{13} = -60, w_{23} = 60,$$

and

$$b_1 = -90, b_2 = -40, b_3 = -30.$$

Use the sigmoid function, $\sigma(v)$, in your program.

3. Use backpropagation to update the weights w_{11}, w_{12}, w_{21} and w_{22} for the XOR gate ANN in Example 8.3.1.

4. Download the "Boston-Housing-ANN-Calculator.ipynb" notebook from GitHub:

`https://github.com/proflynch/A-Simple-Introduction-to-Python.`

Run the cell in the notebook to create an ANN which will value the houses in Boston's suburbs in 1978. Three attributes are used in the program (average number of rooms, index of accessible radial highways, percentage lower status of population). The target data is median value of owner-occupied homes.

(a) Investigate how the learning rate η and the number of epochs affects the results.

(b) List 10 attributes which would be important to you when purchasing your house.

An Introduction to Data Science

Data Science is the field of study combining programming, statistics, mathematics and domain expertise to work with data.

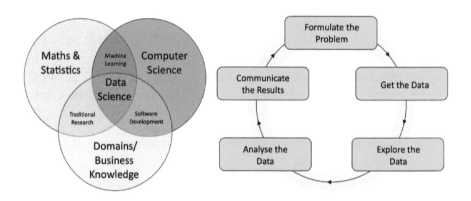

The data science cycle on the right shows a typical workflow algorithm.

9.1 INTRODUCTION TO PANDAS

Python and Data Analysis (PANDAS) is an open source Python package used to solve real-world problems. More information can be found here:

https://pandas.pydata.org.

DOI: 10.1201/9781003472759-9

Example 9.1.1. Create a data frame of student satisfaction at a gym. Out[1] lists the data frame, named df, in this case. Note that the **Dates** are set up to act as an index.

Date	Day	Member Numbers	Staff Numbers	Member/Staff Ratio	Satisfaction (%)
2024-01-08	Sunday	126	12	10.5	44.0
2024-01-09	Monday	34	6	5.7	82.0
2024-01-10	Tuesday	42	6	7.0	74.0
2024-01-11	Wednesday	100	12	8.3	66.0
2024-01-12	Thursday	54	6	9.0	85.0
2024-01-13	Friday	41	6	6.8	88.0
2024-01-14	Saturday	105	12	8.8	45.0

```
In [1]: # A Data Frame of Membership Satisfaction at a Gym.
import numpy as np
import pandas as pd
df=pd.DataFrame({
    "Date": pd.Categorical(["2024-01-08","2024-01-09",
                            "2024-01-10","2024-01-11",
                            "2024-01-12","2024-01-13",
                            "2024-01-14"]),
    "Day " : pd.Categorical(["Sunday","Monday","Tuesday",
                            "Wednesday","Thursday","Friday",
                            "Saturday"]),
    "Member Numbers" : pd.Series([126,34,42,100,54,41,105],
                dtype="int32"),
    "Staff Numbers" : pd.Series([12,6,6,12,6,6,12],
                dtype="int32"),
    "Member/Staff Ratio" : \
    pd.Series([10.5,5.7,7.0,8.3,9.0,6.8,8.8],\
                                    dtype="float32"),
    "Satisfaction (%)" : \
    pd.Series([44,82,74,66,85,88,45],dtype="float32")
    })
df = df.set_index("Date")
df
```

```
In [2]: # List the data types of the columns.
df.dtypes
```

```
Out [2]:
Day                     category
Member Numbers             int32
Staff Numbers              int32
Member/Staff Ratio       float32
Satisfaction (%)         float32
dtype: object
```

```
In [3]: # Write the data to an excel file.
df.to_excel("Gym_Satisfaction_Ratings.xlsx")
```

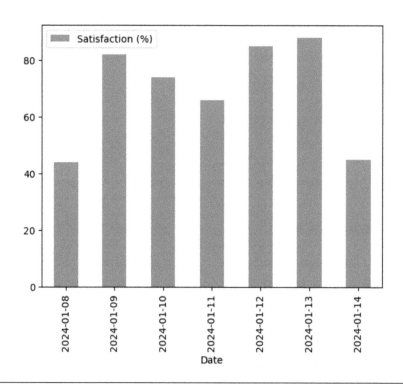

```
In [4]: # Plot a bar chart of the satisfaction ratings.
import matplotlib.pyplot as plt
df.plot.bar(y = "Satisfaction (%)")
```

9.2 LOAD, CLEAN AND PREPROCESS THE DATA

In this section we work with real-world data. We use Data-1-OCR.xlsx, used by the OCR examining body for A-Levels in the UK. Readers can download the Excel file from GitHub:

https://github.com/proflynch/A-Simple-Introduction-to-Python.

The Large Data Set (LDS) consists of data about countries from the CIA World Factbook and from the World Bank. By loading the data using Pandas and setting data frame (df), all columns containing numerical data with commas (data type object) are replaced with floating point numbers. So Pandas pre-processes the data for you. The command df.head() prints the first five rows of the data frame df.

```
In [4]: # Load the data and print the first five rows and
# certain columns.
import pandas as pd      # Import pandas for data analysis.
import seaborn as sns    # Import seaborn for visualisations.
import matplotlib.pyplot as plt
df = pd.read_excel("Data-1-OCR.xlsx" , sheet_name = "Data")
# df.head() # To see the first 5 rows of all data.
df.iloc[0 : 5 , [0 , 1 , 3 , 14 , 19]]
```

	no	Country	population	Life expectancy at birth 1960	Life expectancy at birth 2010
0	1	Algeria	41657488	46.138	74.676
1	2	Egypt	99413317	48.056	70.357
2	3	Libya	6754507	42.609	71.643
3	4	Morocco	34314130	48.458	73.999
4	5	Sudan	43120843	48.194	62.620

Adding a New Numeric Feature To explore the change in life expectancy you could add a feature that calculates the change in life expectancy. Run the code below to create a new feature "Life expectancy change 1960 to 2010," by subtracting the life expectancies in 1960 from those in 2010.

```
In [5]: # Create a new numeric column.
df["Life expectancy change 1960 to 2010"] = \
df["Life expectancy at birth 2010"] - \
df["Life expectancy at birth 1960"]
# Check the data.
df.iloc[0 : 5 , [0 , 1 , 3 , 14 , 19 , 20]]
```

	no	Country	population	Life expectancy at birth 1960	Life expectancy at birth 2010	Life expectancy change 1960 to 2010
0	1	Algeria	41657488	46.138	74.676	28.538
1	2	Egypt	99413317	48.056	70.357	22.301
2	3	Libya	6754507	42.609	71.643	29.034
3	4	Morocco	34314130	48.458	73.999	25.541
4	5	Sudan	43120843	48.194	62.620	14.426

Adding a New Categorical Feature The World Bank defines a High Income Economy as one having a GDP per capita greater than 12,376 dollars. Run the code below to create a new categorical feature "Income Category" which describes whether a country is a high income economy.

```
In [6]: # If the GDP > $12,376 then High, otherwise Low.
df["Income category"] = (df["GDP per capita (US$)"] > \
                    12376).map({True: "High", \
                         False: "Low"})
# List the data frame with certain columns.
df.iloc[0 : 5 , [0, 1, 3, 9, 20]]
```

	no	Country	population	GDP per capita (US$)	Income category
0	1	Algeria	41657488	15200.0	High
1	2	Egypt	99413317	12700.0	High
2	3	Libya	6754507	9600.0	Low
3	4	Morocco	34314130	8600.0	Low
4	5	Sudan	43120843	4300.0	Low

9.3 EXPLORING THE DATA

You can explore the data by finding the dimensions of the data set with **shape** and displaying the data types with the **info()**.

```
In [7]: df.shape # The dimensions of the DataFrame.
```

```
Out[7] (236, 22)
```

```
In [8]: df.info() # Information about the DataFrame.
```

Out[8]:

```
<class 'pandas.core.frame.DataFrame'>
RangeIndex: 236 entries, 0 to 235
Data columns (total 22 columns):
 #   Column                                       Non-Null Count  Dtype
---  ------                                       --------------  -----
 0   no                                           236 non-null    int64
 1   Country                                      236 non-null    object
 2   Region                                       236 non-null    object
 3   population                                   236 non-null    int64
 4   birth rate per 1000                          226 non-null    float64
 5   death rate per 1000                          226 non-null    float64
 6   median age                                   228 non-null    float64
 7   labor force                                  231 non-null    float64
 8   unemployment (%)                             218 non-null    float64
 9   GDP per capita (US$)                         228 non-null    float64
 10  physician density (physicians/1000 population) 198 non-null  float64
 11  Health expenditure (% of GDP)                192 non-null    float64
 12  Total area                                   236 non-null    float64
 13  Land borders                                 236 non-null    object
 14  Life expectancy at birth 1960                188 non-null    float64
 15  Life expectancy at birth 1970                190 non-null    float64
 16  Life expectancy at birth 1980                192 non-null    float64
 17  Life expectancy at birth 1990                195 non-null    float64
 18  Life expectancy at birth 2000                199 non-null    float64
 19  Life expectancy at birth 2010                198 non-null    float64
 20  Life expectancy change 1960 to 2010          188 non-null    float64
 21  Income category                              236 non-null    object
dtypes: float64(16), int64(2), object(4)
memory usage: 40.7+ KB
```

Example 9.3.1. Create box and whisker plots of life expectancy at birth in 2010, grouped by region.

```
In [9]: sns.catplot(data=df, kind="box", x="Life expectancy \
           at birth 2010",  y="Region", aspect=2)
```

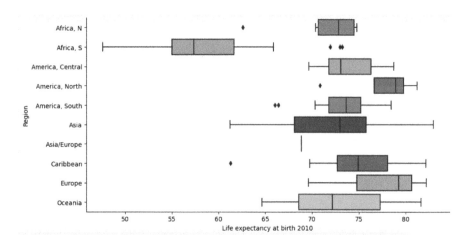

Out[9]: # Box and whisker plots.

Communicate the Results: The life expectancy in South Africa is much worse than all other regions. The best life expectancies are in Europe and North America. Data Scientists would want to know why the life expectancies are so low in South Africa and use data and intervention to try to improve the results.

9.4   VIOLIN, SCATTER AND HEXAGONAL BIN PLOTS

A violin plot plays a similar role as a box and whisker plot. It shows the distribution of quantitative data across several levels of one (or more) categorical variables such that those distributions can be compared. Unlike a box plot, in which all of the plot components correspond to actual datapoints, the violin plot features a kernel density estimation of the underlying distribution. This can be an effective and attractive way to show multiple distributions of data at once, but keep in mind that the estimation procedure is influenced by the sample size, and violins for relatively small samples might look misleadingly smooth.

Example 9.4.1. Create violin plots of life expectancy at birth in 2010, grouped by region.

```
In [10]: sns.violinplot(data=df, x="Life expectancy at \
             birth 2010", y="Region")
```

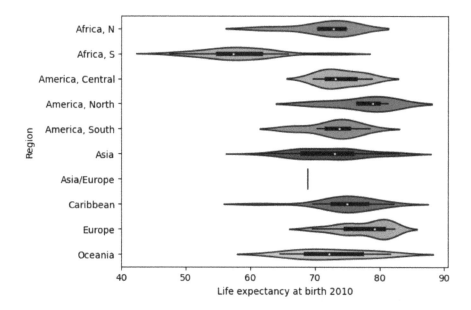

Figure 9.1 Out[10]: Violin plots, compare with the box and whisker plots earlier.

Figure 9.1 shows the violin plots.

Scatter plots are useful to see correlations between two variables. The coordinates of each point are defined by two DataFrame columns and filled circles are used to represent each point. In the example below, the points are color-coded by region. The **aspect** function changes the ratio of the width to the height within the chart arguments.

Example 9.4.2. Create a scatter plot of GDP per capita (US dollars) against life expectancy at birth in 2010, color-coded by region.

```
In [11]: sns.relplot(data=df, x="GDP per capita (US$)", \
            y="Life expectancy at birth 2010", \
            hue="Region", aspect=1.5)
```

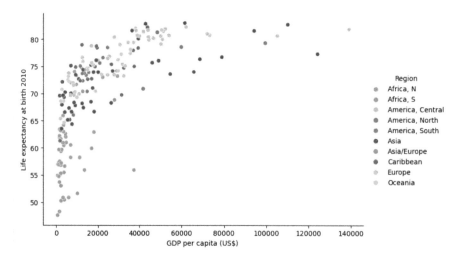

Figure 9.2 Out[11]: Scatter plot color-coded by region. The life expectancy at birth in South Africa is much worse than all other regions.

Figure 9.2 shows Out[11].

Hexagonal Bin Plots: When you have a very large data set, with thousands or even millions of data points, scatter plots can be misleading. This is because the points start landing on top of each other, making it hard to tell if there are many points in an area or just a few. To solve this, some charts 'bin' the data first, in much the same way a histogram does. Unlike a histogram, a bin plot shows density using shading with colors representing higher densities. Seaborn can create a bin plot using its **jointplot function**, which also provides histograms of the two variables. Run the code below to create a hexagonal bin plot of life expectancy against GDP.

Example 9.4.3. Plot a hexagonal plot of GDP per capita (US dollars) against life expectancy at birth 2010.

```
In [12]: sns.jointplot(data=df, kind="hex", x="GDP per \
         capita (US$)", y="Life expectancy at birth 2010")
```

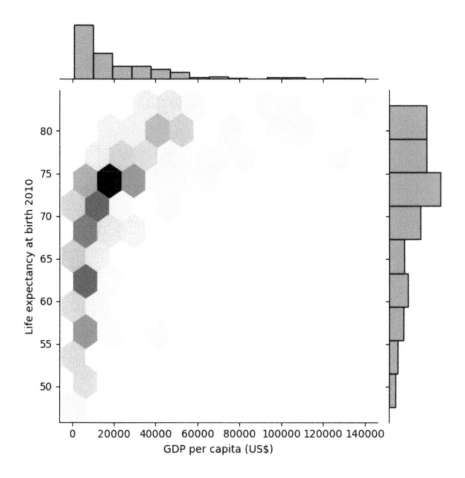

Figure 9.3 Out[12]: Hexagonal bin plot of GDP per capita (US dollars) against life expectancy at birth 2010.

Figure 9.3 shows Out[12].

EXERCISES

The data sets used below are taken from the A-Level Mathematics data sets used in the UK. All data sets can be downloaded through GitHib: https://github.com/proflynch/A-Simple-Introduction-to-Python.

1. For the data set Data-1-OCR.xlsx, create a scatter plot of GDP against life expectancy at birth in 2010, where the size of the filled circles is determined by physician density. What can you conclude?

2. Load the data file, Data-2-Edexcel.csv from GitHub. This LDS consists of weather data samples provided by the UK Met Office for five UK weather stations. Load the data and set up a data frame. Explore the data. Compare the daily mean temperature at the five stations in 1987 compared to 2015. Communicate the results.

3. Load the data file, Data-3-AQA.xlsx from GitHub. This LDS has been taken from the UK Department for Transport Stock Vehicle Database. Load the data and set up a data frame. Explore the data. Create a scatter diagram for carbon dioxide emissions against mass for the petrol cars only. Communicate the results.

4. Load the data file, Data-4-OCR.xlsx from GitHub. This LDS consists of four sets of data: two each from the censuses of 2001 and 2011; two on methods of travel to work and two showing the age structure of the population. Load the data and set up a data frame. Explore the data. Add a column giving the percentage of people in employment who cycle to work. Produce a box and whisker plot of percentage of workers who cycle to work against region. Communicate the results.

An Introduction to Object Oriented Programming

Object-Oriented Programming (OOP) is a programming paradigm based on the concept of "object," which can contain data and code. The data is in the form of attributes, and the code is in the form of methods.

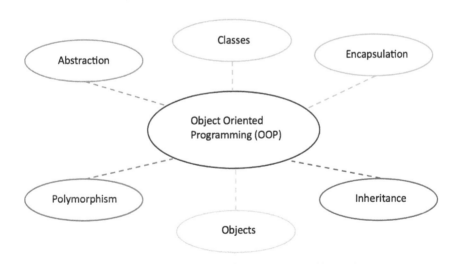

10.1   CLASSES AND OBJECTS

Attributes are the characteristics of the class and methods are the actions that can be performed on the class.

Example 10.1. Set up a Student class with attributes, methods and objects of your choice.

```
In [1]: # Define a class.
class Student:
    # Attributes.
    def __init__(self, name, age, course, student_id):
        self.name = name
        self.age = age
        self.course = course
        self.student_id = student_id
    # Methods.
    def enrol(self):
        return "Enrolled"
    def payfees(self):
        return "Fees Paid"
    def accommodation(self):
        return "Student in Accommodation"
# Create objects of the Student class.
Ali_M = Student("Mohammad Ali", 18, "Software Engineering", \
 23000001)
Parker_P = Student("Peter Parker", 19, "Physics", 23000002)
Kane_K = Student("Kate Kane", 22, "Biology", 23000003)
Chang_A = Student("Angela Chang", 18, "Music", 23000004)
```

```
In [2]: # Using attributes.
print(Ali_M.name)
print(Parker_P.age)
print(Kane_K.course)
print(Chang_A.student_id)
```

Out [2]: Mohammad Ali
19
Biology

23000004

```
In [3]: # Using methods.
print(Ali_M.enrol())
print(Parker_P.payfees())
print(Kane_K.accommodation())
```

Out[3]: Enrolled
Fees Paid
Student in Accommodation

10.2   ENCAPSULATION

Encapsulation is one of the core concepts in object-oriented programming and describes the attributes and methods operating on this data
into one unit.

Advantages of Encapsulation:

1. Security: The main advantage of using encapsulation is in the se-
   curing of data. Encapsulation protects an object from unauthorized
   access. It allows private and protected access levels to prevent ac-
   cidental data modification.

2. Data Hiding: The user would not be aware of what is going on be-
   hind the scenes. To modify a data member, call the setter method.
   To read a data member, call the getter method. What these setter
   and getter methods are doing is hidden from the user.

3. Simplicity: It simplifies the maintenance of the application by keep-
   ing classes separated and preventing them from tightly coupling
   with each other.

4. Aesthetics: Bundling attributes and methods within a class makes
   code more readable and maintainable.

```
In [4]: # Encapsulation and data hiding.
class Employee:
    # Attributes
    def __init__(self, name, job_title, salary):
```

```
            # Public attributes.
            self.name = name
            self.job_title = job_title
            # Private attribute.
            self.__salary = salary # Data hiding.
        # Methods
        def work(self):
            print(self.name, "is employed in:", self.job_title)
# Creating objects of a class
emp1 = Employee("Jessa", "Human Resources", 25000)
emp2 = Employee("Tom", "Lecturing", 35000)
emp3 = Employee("Leon", "Administration", 15000)
emp1.work()
emp2.work()
emp3.work()
```

Out[4]: Jessa is employed in: Human Resources
Tom is employed in: Lecturing
Leon is employed in: Administration

```
In [5]: # Accessing private data members.
print(emp1.name)
print(emp1.job_title)
print(emp1.__salary)
```

Out[5]: Jessa
Human Resources
AttributeError: 'Employee' object has no attribute '__salary'

10.3   INHERITANCE

The process of inheriting the properties of the parent class into a child class is called inheritance. The existing class is called a base class or parent class and the new class is called a subclass or child class or derived class.

```
In [6]: # Parent class.
class Person:
    def __init__(self, name, age):
```

```
        print("Inside Person class")
        print("Name:", name, "Age:", age)
# Parent class 2.
class Company:
    def __init__(self, company_name, location):
        print("Inside Company class")
        print("Name:", company_name, "Location:", location)
# Child class.
class Employee(Person, Company):
    def __init__(self, department, job_title):
        super().__init__(department, job_title)
        print("Inside Employee class")
        print("Department:", department, "Job_Title:", \
            job_title)
# Create object of Employee
emp1 = Person("Jessa" , 35)
emp1 = Employee("Software", "Machine Learning")
emp1 = Company("Google", "Atlanta")
```

```
Out[6]: Inside Person class
Name: Jessa Age: 35
Inside Person class
Name: Software Age: Machine Learning
Inside Employee class
Department: Software Job_Title: Machine Learning
Inside Company class
Name: Google Location: Atlanta
```

10.4 POLYMORPHISM

Polymorphism is one of the core concepts of OOP and describes situations in which something occurs in several different forms.

Function Polymorphism: An example of a Python polymorphic function that can be used on different objects is the len() function, which can be used on lists, sets, strings, tuples and dictionaries.

```
In [7]: # A list.
```

```
print(len([1 , 2 , 3 , 4 , 5 , 6 , 7 , 8]))
# A set.
print(len({1 , 2 , "apple" , "cherry"}))
# A string.
print(len("This is a string"))
# A tuple.
print(len(("apple", "banana", "cherry", "pear", "orange")))
# The final example is a dictionary.
print(len({"Manufacturer": "Ford", "Model" : "Mustang" , \
          "Year" : 1964}))
```

```
Out[7]: 8
4
16
5
3
```

Inheritance Class Polymorphism

What about classes with child classes with the same name? Can we use polymorphism there?

Yes. If we use the example below and make a parent class called Vehicle, and make Car, Boat, Plane child classes of Vehicle, the child classes inherits the Vehicle methods, but can override them.

Example 10.4.1. Demonstrate inheritance class polymorphism using vehicles of your choice.

```
In [8]: # Inheritance Class Polymorphism.
class Vehicle:
    # Attributes.
    def __init__(self, manufacturer, model):
        self.brand = manufacturer
        self.model = model
    # Method.
    def move(self):
        print("Move!")
class Car(Vehicle):
```

```
    pass
class Boat(Vehicle):
    def move(self):
        print("Sail!")
class Plane(Vehicle):
    def move(self):
        print("Fly!")
car_1 = Car("Kia", "Sportage")        #Create a Car object.
boat_1 = Boat("Yamaha", "SX 210")     #Create a Boat object.
plane_1 = Plane("Boeing", "747")      #Create a Plane object.
for x in (car_1, boat_1, plane_1):
    print(x.brand)
    print(x.model)
    x.move()
```

```
Out[8]: Kia
Sportage
Move!
Yamaha
SX 210
Sail!
Boeing
747
Fly!
```

EXERCISES

1. You have been employed by Chester Zoo, UK, as a software engineer. Create an Animal class with attributes name, age, sex and feeding time. The Animal class has three methods, resting, moving and sleeping.

2. Create ten animal objects of your choice and introduce private attributes called "feed cost" and "vet cost."

3. The following program shows a parent class **Pet** and child classes **Cat** and **Canary**. The objects **Tom** and **Cuckoo** have been declared.

```
In [9]: # Parent and child classes.
class Pet:
    def __init__(self , legs):
        self.legs = legs
    def walk(self):
        print("Pet parent class. Walking...")
class Cat(Pet):
    def __init__(self , legs , tail):
        self.legs = legs
        self.tail = tail
    def meeow(self):
        print("Cat child class.")
        print("A cat meeows but a canary can't. Meeow...")
class Canary(Pet):
    def chirp(self):
        print("Canary child class.")
        print("A canary chirps but a cat can't. Chirp...")
Tom = Cat(4 , True)
Cuckoo = Canary(2)
```

What output do the following lines give?

```
In [10]: print(Tom.legs)
print(Tom.tail)
print(Cuckoo.legs)
Tom.meeow()
Cuckoo.chirp()
```

4. The figure below is a screenshot from the **Brick Breaker Game.**
The player moves the paddle left and right to hit the ball which
rebounds off the paddle, bricks and three perimeter walls. If the
ball passes the paddle, the player loses a life. The aqua colored
bricks (lowest level) break after one hit, the tomato colored bricks
(middle level) break after two hits, and the lawn green colored
bricks (top level) break after three hits. You win the game if all
bricks are destroyed.

(a) List at least five objects, attributes and methods, respectively,
required to create a Brick Breaker Game.

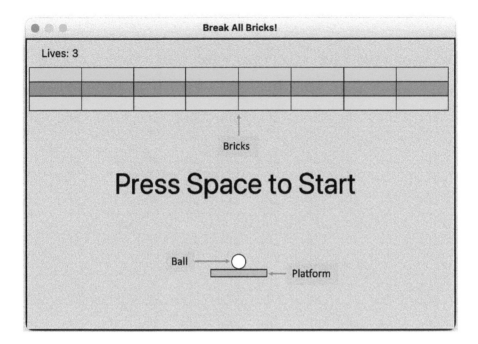

(b) Download the "Brick-Breaker-Game.ipynb" notebook from GitHub:

https://github.com/proflynch/A-Simple-Introduction-to-Python.

Run the cell in the notebook to play the Brick Breaker Game. Can you destroy all of the bricks?

(c) What other physical quantities could be modeled to make the game more realistic?

Index